环境国际公约履约系列丛书

生态环境部对外合作与交流中心研究成果

联合国工业发展组织项目成果

UNITED NATIONS
INDUSTRIAL DEVELOPMENT ORGANIZATION

新冠肺炎疫情期间
中国医疗废物管理培训指南

生态环境部对外合作与交流中心
国家医疗废物管理培训基地（上海） 主编

中国环境出版集团 · 北京

图书在版编目（CIP）数据

新冠肺炎疫情期间中国医疗废物管理培训指南 / 生态环境部对外合作与交流中心，国家医疗废物管理培训基地（上海）主编. -- 北京 ：中国环境出版集团，2020.11（2021.2重印）

（环境国际公约履约系列丛书）

ISBN 978-7-5111-4494-2

Ⅰ．①新… Ⅱ．①生… ②国… Ⅲ．①日冕形病毒－病毒病－肺炎－医用废弃物－废物管理－技术培训－指南 Ⅳ．①X799.5-62

中国版本图书馆CIP数据核字（2020）第217988号

出 版 人 武德凯
责任编辑 曹 玮
责任校对 任 丽
封面设计 岳 帅

出版发行 中国环境出版集团
（100062 北京市东城区广渠路内大街 16 号）
网 址：http://www.cesp.cn
电子邮箱：bjgl@cesp.com.cn
联系电话：010-67112765（编辑管理部）
发行热线：010-67125803,010-67113405（传真）

印　　刷 北京建宏印刷有限公司
经　　销 各地新华书店
版　　次 2020 年 11 月第 1 版
印　　次 2021 年 2 月第 2 次印刷
开　　本 880×1230 1/32
印　　张 4.875
字　　数 76 千字
定　　价 30.00 元

编委会

Editorial Committee

前 言
Foreword

医疗废物是指医疗卫生机构在医疗、预防、保健及其他相关活动中，产生的具有直接或者间接感染性、毒性及其他危害性的废物。做好医疗废物安全管理工作，对于提供高质量、以人为本的医疗服务，切实维护人民群众身体健康和环境安全，加快推进全球生态文明建设，促进经济社会可持续发展具有重大意义。

2020 年，新型冠状病毒感染的肺炎疫情在全球暴发蔓延，受疫情影响国家的医疗废物产生量激增。如何确保疫情期间医疗废物得到及时、安全、有序、高效的处理处置，这给各国政府提出了极大的挑战。新冠病毒具有极高感染性，确保高感染性的相关医疗废物妥善处置、有效阻断病毒传播的二次污染，对取得抗击疫情的最终胜利至关重要。

为进一步加强疫情期间医疗废物的规范化管理和安全处置，在商务部的协调和支持下，生态环境部（对外合作与交流中心作为代表）同联合国工业发展组织（UNIDO）紧急合作开发并实施了新型冠状病毒疫情中国应急响应援助项目。在生态环境部指导下，对外合作与交流中心依托项目与国家医疗废物管理培训基地（由上海市固体废物管理中心和上海市固体废物处置有限公司组建）共同编制完成了《新冠肺炎疫情期间中国医疗废物管理培训指南》，并得到了生态环境部固体废物与化学品管理技术中心、巴塞尔公约亚太区域中心（清华大学环境学院）、中国科学院大学、河南省生态环境厅等单位有关专家的大力支持。

本指南立足于中国现行的医疗废物管理体系，紧密结合工作实际，高度结合疫情期间中国在医疗废物管理与应急方面的一系列举措，涵盖医疗废物概况、分类收集、包装，医疗卫生机构内部运送，医疗废物暂时贮存、转移、处置，医疗废物处理过程中的 EHS 管理，医疗废物监督管理，疫情期间医疗废物管理规定、医疗废物收集暂存、医疗废物处置及医疗废物应急处置案例等内容，汇集了上海、武汉等城市在新冠肺炎疫情防控工作中医疗废物应急管理与处置的经验和智慧。

病毒没有国界，疫情不分种族，人类是休戚与共的命运共同体，唯有团结协作、携手应对，国际社会才能最终战胜疫情。团结合作是国际社会战胜疫情最有力的武器，为坚决打好新冠肺炎疫情防控全球阻击战，中国愿将疫情防控的有益做法同各国分享，并向出现疫情扩散的国家提供力所能及的援助。为促进将中国有效应对疫情医疗废物管理的经验转化为其他有需要国家抗击疫情的有力武器，生态环境部对外合作与交流中心将该书编译为中英文版。在这场全球战“疫”中，相信本书对人类最终战胜疫情将起到积极作用。

本书的编写是一项探索性工作，全体编审人员为此付出了很多艰辛和努力，同时也得到了相关领导和专家的悉心指导，在此一并表示感谢。由于时间紧迫，不足之处在所难免，欢迎提出宝贵意见和建议。

编写组

2020 年 5 月

目录

CONTENTS

第一部分 医疗废物基本概念

第二部分　医疗废物全过程管理

Part 1

第一部分

医疗废物基本概念

1 定义及概况

1.1 医疗废物的定义是什么？

根据《医疗废物管理条例》，医疗废物是指医疗卫生机构在医疗、预防、保健以及其他相关活动中所产生的具有直接或间接感染性、毒性以及其他危害性的废物。医疗卫生机构收治的传染病病人或者疑似传染病病人产生的生活垃圾，按照医疗废物进行管理和处置。另外，计划生育技术服务，医学科研、教学，尸体检查和其他相关活动中产生的具有直接或者间接感染性、毒性以及其他危害性废物的管理，依照《医疗废物管理条例》执行。医疗卫生机构废弃的麻醉、精神、放射性、毒性等药品及其相关的废物管理，依照有关法律、行政法规和国家有关规定、标准执行。

1.2 医疗废物可能造成哪些危害？

（1）感染性废物、病理性废物和医用锐器可能引起感染性疾病，甚至引起传染病流行，如乙型肝炎、丙型肝炎、艾滋病等。

（2）遗传毒性废物可能引起基因突变导致遗传性疾病。

（3）药物性废物和化学性废物可能引起药物中毒和化学性损害。

（4）压力容器和锐器可能导致物理性损害。

1.3 医疗废物危害的对象有哪些？

所有接触（暴露、直接和间接接触）医疗废物的人将可能受到潜在威胁，包括产生医疗废物的医疗卫生机构工作人员，探视者，家庭医疗保健时医疗废物的接触者，医疗废物收集、贮存、转运、处置工作人员，医疗废物环境监管和卫生防疫监管人员，由于医疗废物管理不善而导致暴露的人群，以及医疗废物处置产生二次污染的受众。

1.4 如何对医疗废物进行分类?

根据《医疗废物分类目录》，医疗废物分为 5 大类：感染性废物、病理性废物、损伤性废物、药物性废物、化学性废物。各类医疗废物特征及常见组分或医疗废物名称见表 1-1。

表 1-1　各类医疗废物特征及常见组分或医疗废物名称

类别	特征	常见组分或医疗废物名称	示意图
感染性废物	携带病原微生物，具有引发感染性疾病传播危险的医疗废物	被病人血液、体液、排泄物污染的物品包括：棉球、棉签、引流棉条、纱布及其他各种敷料；一次性使用卫生用品、一次性使用医疗用品及一次性医疗器械；废弃的被服；其他被病人血液、体液、排泄物污染的物品	
		医疗卫生机构收治的隔离传染病病人或者疑似传染病病人产生的生活垃圾	
		病原体的培养基、标本和菌种、毒种保存液	
		各种废弃的医学标本	
		废弃的血液、血清	
		使用后的一次性使用医疗用品及一次性医疗器械视为感染性废物	

类别	特征	常见组分或医疗废物名称	示意图
病理性废物	诊疗过程中产生的人体废弃物和医学实验动物尸体等	手术及其他诊疗过程中产生的废弃的人体组织、器官等	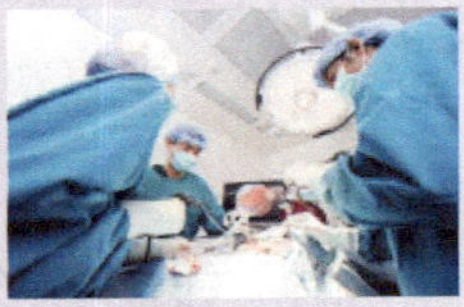
		医学实验动物的组织、尸体	
		病理切片后废弃的人体组织、病理蜡块等	
损伤性废物	能够刺伤或者割伤人体的废弃的医用锐器	医用针头、缝合针	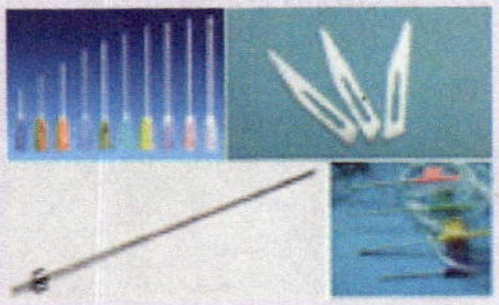
		各类医用锐器，包括解剖刀、手术刀、备皮刀、手术锯等	
		载玻片、玻璃试管、玻璃安瓿等	
药物性废物	过期、淘汰、变质或者被污染的废弃的药品	废弃的一般性药品，如抗生素、非处方类药品等	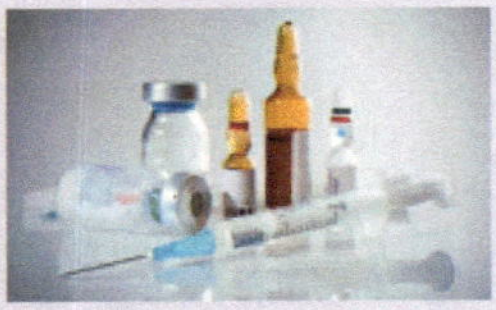
		废弃的细胞毒性药物和遗传毒性药物，包括：致癌性药物，如硫唑嘌呤、苯丁酸氮芥、萘氮芥、环孢霉素、环磷酰胺、苯丙氨酸氮芥、司莫司汀、三苯氧氨、硫替派等；可疑致癌性药物，如顺铂、丝裂霉素、阿霉素、苯巴比妥等；免疫抑制剂	
		废弃的疫苗、血液制品等	
化学性废物	具有毒性、腐蚀性、易燃易爆性的废弃的化学物品	医学影像室、实验室废弃的化学试剂	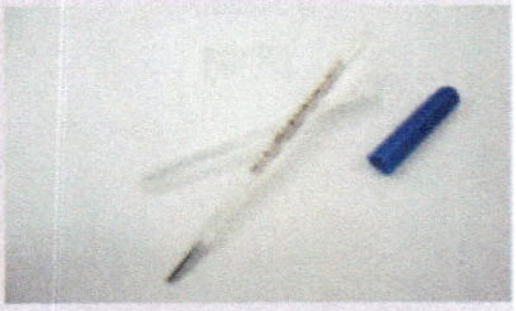
		废弃的过氧乙酸、戊二醛等化学消毒剂	
		废弃的汞血压计、汞温度计	

1.5 我国医疗废物管理相关的法律、法规有哪些？

我国医疗废物管理主要依据2003年颁布的《医疗废物管理条例》，该条例是根据《中华人民共和国传染病防治法》和《中华人民共和国固体废物污染环境防治法》两部上位法制定的，适用于医疗废物的收集、运送、贮存、处置以及监督管理等活动，该条例规定卫生部门、环保部门为两个主要监管部门——卫生部门负责监督管理全过程的疾病防治、环保部门负责监督管理全过程的环境污染防治。在该条例出台之前，环保部门就已将“医院临床废物”纳入《国家危险废物名录》（1998年版），从而将医疗废物包含在《中华人民共和国固体废物污染环境防治法》《危险废物经营许可证管理办法》《危险废物转移联单管理办法》等组成的危险废物管理规则体系内，涉及危险废物贮存污染控制、危险货物运输管理要求，以及危险废物处置工程技术规范、污染控制和排放限值标准等一系列标准。该条例出台之后，卫生部门、环保部门也专门制定了医疗废物法规及管理政策（但仍包含在危险废物管理框架内），涵盖医疗废物分类、处理处置技术规范、行政处罚等。主要的专门法规还包括：《医疗废物分类目录》《医疗卫生机构医疗废物管理办法》《医疗废物管理行政处罚办法》。

疫情期间的医疗废物管理，除执行日常管理要求外，还应满足特殊规定。这些规定主要包括《医疗废物集中处置技术规范（试行）》（第六章　重大传染病疫情期间医疗废物处置）、《“SARS”病毒污染的废弃物应急处理处置技术方案》、《应对甲型H1N1流感疫情医疗废物管理预案》、《新型冠状病毒感染的肺炎疫情医疗废物应急处置管理与技术指南（试行）》、《关于做好新型冠状病毒感染的肺炎疫情期间医疗卫生机构医疗废物管理工作的通知》。

相关法律、法规、技术规范详见表 1-2。

表 1-2 医疗废物管理相关的法律、法规等

序号	主题	名称	年份
1	上位法	中华人民共和国传染病防治法	2004
2		中华人民共和国固体废物污染环境防治法	2020
3	专门法规	医疗废物管理条例	2011
4		医疗废物分类目录	2003
5		医疗卫生机构医疗废物管理办法	2003
6		医疗废物管理行政处罚办法	2010
7	总体要求	医疗废物集中处置技术规范（试行）	2003
8	包装	医疗废物专用包装袋、容器和警示标志标准（HJ 421—2008）	2008
9	运输	道路危险货物运输管理规定	2013
10		医疗废物转运车技术要求（试行）（GB 19217—2003）	2003
11	处置技术	医疗废物集中焚烧处置工程技术规范（HJ/T 177—2005）	2005
12		医疗废物化学消毒集中处理工程技术规范（试行）（HJ/T 228—2006）	2006
13		医疗废物微波消毒集中处理工程技术规范（试行）（HJ/T 229—2006）	2006
14		医疗废物高温蒸汽集中处理工程技术规范（试行）（HJ/T 276—2006）	2006
15		医疗废物集中焚烧处置设施运行监督管理技术规范（试行）（HJ 516—2009）	2010
16		医疗废物处理处置污染防治最佳可行技术指南（试行）（HJ-BAT-8）	2012
17		医疗废物焚烧炉技术要求（试行）（GB 19218—2003）	2003
18	行政管理	国家危险废物名录	2016
19		危险废物经营许可证管理办法	2004
20		危险废物转移联单管理办法	1999
21		危险废物和医疗废物处置设施建设项目环境影响评价技术原则（试行）	2014

序号	主题	名称	年份
22	行政管理	医疗废物集中焚烧处置设施运行监督管理技术规范（试行）	2010
23	疫情医疗废物管理	医疗废物集中处置技术规范（试行）	2003
24		“SARS”病毒污染的废弃物应急处理处置技术方案	2003
25		应对甲型 H1N1 流感疫情医疗废物管理预案	2009
26		新型冠状病毒感染的肺炎疫情医疗废物应急处置管理与技术指南（试行）	2020
27		关于做好新型冠状病毒感染的肺炎疫情期间医疗卫生机构医疗废物管理工作的通知	2020

1.6 世界卫生组织对医疗废物有什么管理要求?

1998 年，世界卫生组织（WHO）发布报告，从国家立法和管理行动计划，医疗卫生机构内部管理，医疗废物减量措施，医疗废物的收集、分类、包装、运输、处理、处置，优先支持的活动，以及发展战略等方面对医疗废物管理提出了相应的要求。报告提出，安全可靠的医疗废物处置方法是最重要的；进行有效的医疗废物管理和处置客观上要求全社会的合作和协调；建立一整套法律法规体系、开展人员培训、提高公共意识是成功实施医疗废物管理和处置的重要条件。

2014 年，WHO 发布修订版的《医疗废物无害化管理手册》，对在常规和应急情况下医疗废物的管理进行了指导。2014 年 12 月，WHO 针对埃博拉病毒发布一系列指南文件，对医疗废物的简易处理处置给出了指导意见。其中对发生自然灾害或战争冲突的应急状态下的医疗废物处理处置做出规定，在资源不可用时，医疗废物可采用以下最低限度的处理和处置方法：在坑或壕沟中进行现场掩埋、在简易双室焚烧炉中焚烧、将尖锐废物或少量药品封装起来然后就地掩埋或埋在市政倾倒区的特殊区域内、在高温工业焚化炉中焚烧（前提是有安全的运输工具）、用小型高压灭菌器对感染性废物和尖锐废物进行消毒等。2015 年 6 月，WHO 发布《中东呼吸综合征冠状病毒（MERS-CoV）住院患者的感染预防和控制临时建议》，2020 年 1 月 25 日 WHO 发布《疑似新型冠状病毒（COVID-19）感染时的卫生保健中的感染预防和控制临时导则》，均指出“医疗废物应按照常规程序进行管理”。

1.7 《关于持久性有机污染物的斯德哥尔摩公约》对医疗废物有什么管理要求？

二噁英类是《关于持久性有机污染物的斯德哥尔摩公约》要求控制的首要无意产生类持久性有机污染物，医疗废物焚烧是二噁英排放的重要来源。公约要求缔约方采用最佳可行技术和最佳环境实践 (BAT/BEP) 减少二噁英类排放。公约附件 C 中与医疗废物管理相关的 BAT/BEP 一般性措施包括：①减量化；②停止露天焚烧或简易焚烧；③新建设施优先考虑焚烧替代办法；④减少处置过程中的污染排放。

BAT/BEP 导则认为，尽管焚烧技术是医疗废物处置最为成熟的技术，但也具有较大环境风险，鼓励采用高温蒸汽、微波和化学处理等非焚烧替代技术。对医疗废物管理和处置的具体要求是：①推进政府立法；②建立医疗废物分类目录；③推进源头减量，严格区分生活垃圾和医疗废物，减少一次性医疗用品的使用；④减少含氯医疗用品的使用；⑤建立有效的医疗废物管理系统；⑥分类应与处置方式相适应；⑦慎建医疗废物焚烧炉，减少焚烧处置量；⑧在满足公共卫生安全的前提下，尽可能地提倡资源回收利用；⑨鼓励采用先进技术，淘汰落后处置技术；⑩支持开发和推广废物减量化技术研究，推进环境友好的处置技术的研发和应用。

1.8 我国《关于持久性有机污染物的斯德哥尔摩公约》在医疗废物焚烧方面需要做什么？

《关于持久性有机污染物的斯德哥尔摩公约》于 2004 年 11 月 11 日对我国正式生效。二噁英类是公约首批受控物质，公约要求持续减少并在可行的情况下最终消除该类物质。2004 年，医疗废物焚烧占我国二噁英类排放总量的 11.5%，是我国二噁英优先控制的重点行业。按照《中华人民共和国履行〈关于持久性有机污染物的斯德哥尔摩公约〉国家实施计划》要求：到 2008 年，基本建立无意产生 POPs 重点行业有效实施最佳可行技术和最佳环境实践（BAT/BEP）的管理体系，实现对重点行业的新源应用 BAT，促进 BEP；到 2015 年，重点行业广泛开展应用 BAT/BEP，基本控制二噁英类排放增长的趋势；到 2025 年，全面推行 BAT 和 BEP，最大限度消除二噁英类排放。公约秘书处组织编写的 BAT/BEP 导则明确指出，应用 BAT/BEP 后医疗废物焚烧的二噁英类排放浓度将不超过 0.1 ng TEQ/m^3（称为 0.1 标准）。因此，全面实施 0.1 标准是我国医疗废物焚烧履约的最终目标。

1.9 《控制危险废物越境转移及其处置巴塞尔公约》对医疗废物有什么管理要求？

《控制危险废物越境转移及其处置巴塞尔公约》指出，医疗废物具有感染性，是危险废物的重要组成部分之一。该公约将其列入附件I中，核心目标是实现环境无害化管理，保护人体健康，减少其对环境的破坏。为推进该公约实施，公约秘书处根据公约编制了《生物医疗和卫生保健废物环境无害化管理技术准则（Y1；Y3）》。

该准则针对如何避免和防止废物产生，如何实行分类、收集、标识和转运，如何运输和暂存，如何回收利用等提出了相应的要求。在具体处置方法方面，该准则也提出了相应的建议，提出针对不同废物类型所推荐的处置方法，如高温蒸汽处理、化学消毒处理、微波辐照处理、焚烧处置等。

1.10 对医疗卫生机构废弃物有哪些新要求?

2020年2月24日，国家卫生健康委会同生态环境部等10部门联合印发了《医疗卫生机构废弃物综合治理工作方案》（国卫医发〔2020〕3号）。其中强调，一是加强废弃物的分类及源头管理，医疗卫生机构将产生的医疗废物、生活垃圾、输液瓶（袋）等进行分类管理；二是解决医疗废物集中处置设施不足的问题，明确要求未达标省份要在2020年年底前实现每个地级以上城市至少建成1个符合运行要求的医疗废物集中处置设施；三是解决输液瓶（袋）回收利用的问题，明确了“闭环管理、定点定向、全程追溯”的原则，特别是在回收利用环节，由地方出台政策措施，确保辖区内分别至少有1家回收和利用企业或1家回收利用一体化企业，并做到定点定向；四是开展多部门专项整治，在全国范围内开展医疗卫生机构废弃物专项整治行动；五是完善相关保障措施和工作机制，部门之间加强信息沟通并建立协同机制，促进医疗卫生机构产生的各类废弃物及时得到处置；落实医疗卫生机构废弃物的处置政策，合理减轻医疗卫生机构处置的费用负担。

Part 2

第二部分

医疗废物全过程管理

2 医疗废物分类收集

2.1 医疗废物全过程管理包括哪些环节？

医疗废物全过程管理包括医疗卫生机构对医疗废物的分类、收集、包装、运送、暂时贮存，医疗废物的交接，集中处置单位对医疗废物的收运、贮存、处置等环节。

2.2 不同类别的医疗废物是否可以混合收集？

医疗卫生机构应当根据《医疗废物分类目录》，对医疗废物实施分类管理，在医疗废物产生地点设置有医疗废物分类收集方法的示意图或者文字说明，明确感染性废物、病理性废物、损伤性废物、药物性废物及化学性废物不能混合收集。少量的药物性废物可以混入感染性废物，但应当在标签上注明。禁止将医疗废物混入其他废物和生活垃圾。

特别注意：对于高危险废物，如传染病病人产生的医疗废物、排泄物等具有特别要求。部分类别医疗废物分类收集的目的是送往专门机构处置，不能运送至集中处置单位混合处置。医疗卫生机构废弃的麻醉、精神、放射性、毒性等药品及其相关的废物的管理，依照有关法律、行政法规和国家有关规定、标准执行。

2.3 哪些是高危险医疗废物？收集前应如何处理？

《医疗废物管理条例》规定，医疗废物中病原体的培养基、标本和菌种、毒种保存液等高危险废物，应当首先在产生地点进行压力蒸汽灭菌或者化学消毒处理，然后按感染性废物进行收集处理。

2.4 传染病病人产生的生活垃圾应如何收集?

医疗卫生机构收治的传染病病人或者疑似传染病病人产生的生活垃圾，按照医疗废物进行管理和处置。应当使用双层包装物，并及时密封。

2.5 传染病病人的排泄物应如何收集处理?

被隔离的传染病病人或者疑似传染病病人产生的具有传染性的排泄物，应当按照国家规定严格消毒，达到国家规定的排放标准后方可排入污水处理系统。

2.6 化学性医疗废物应如何收集处置?

化学性医疗废物应单独放置于专有的包装容器内，禁止混入其他医疗废物中。易燃、易爆化学性医疗废物应单独收集、运输，医疗卫生机构应按照危险废物处置的有关途径，将其交由具有相应处置能力的危险废物处置单位进行处置，医疗废物集中处置单位应协助落实有关处置去向。

2.7 医疗废物收集的时限是如何要求的?

医疗废物产生单位在本单位内收集医疗废物，应当每天不少于1次；对巡回医疗和现场急救等医疗活动中产生的医疗废物，应当在医疗活动结束后立即完成收集。

2.8 医疗废物分类收集点应符合什么要求?

医疗废物分类收集点是指设在门（急）诊、科室、医技部门、病房病区的用于接收各医疗岗位医务人员送交的医疗废物的暂时贮存点，用于向本机构暂存仓库运送前的医疗废物的分类收集、包装和管理。诊所等小型医疗卫生机构医疗废物分类收集点通常与暂时贮存库、暂存柜合并设置。《上海市医疗废物卫生管理规范》规定，医疗卫生机构内各部门设置的医疗废物分类收集点应当符合以下要求：相对独立，设有相应的分隔设施且易于管理；方便医疗废物的收集、转运；有标明医疗废物分类收集方法的示意图和有关文字说明；放置医疗废物分类收集包装的盛器应当为脚踏开启的封闭硬质盛器。

2.9 家庭病床和医疗服务点的医疗废物由谁负责收集？

在家庭病床和医疗服务点诊疗活动中产生的医疗废物，在每次诊疗活动结束后，由开展诊疗活动的医务人员进行收集，当日能将医疗废物带回医疗卫生机构的，直接交本单位负责医疗废物暂时贮存管理的人员处置，当日不能带回的按要求收集后暂时贮存于病人家中的安全处，次日由负责诊疗活动的医务人员带回。不得交由病人或病人家属自行处置。

2.10 家庭病床和医疗服务点的医疗废物分类收集有什么要求？

家庭病床和医疗服务点产生的医疗废物，其中感染性废物应置于专用包装袋内并封口，系上中文标签，置于专用盛器中；损伤性废物临时置于硬质盛器中并加盖。

2.11　救护车、流动采血车、献血屋的医疗废物如何收集？

对救护、采血活动中产生的感染性废物和损伤性废物，先分别置于专用包装袋内和硬质盛器中，再暂存于车内设置的专用盛器。救护车上暂时贮存的医疗废物应当由随车救护人员交接给救护病人送达的医疗卫生机构急诊室处置；流动采血车、献血屋每天工作完毕后，由专人于当日将暂存的医疗废物交接给所属的采供血机构处置。上述医疗废物转交前应当封口并系上中文标签。

3　医疗废物包装

3.1　医疗废物包装容器是否有专门的标准规范？

《医疗废物专用包装袋、容器和警示标志标准》（HJ 421—2008）规定了医疗废物专用包装袋、利器盒和周转箱（桶）的技术要求以及相应的试验方法和检验规则，并规定了医疗废物警示标志（图 3-1）。

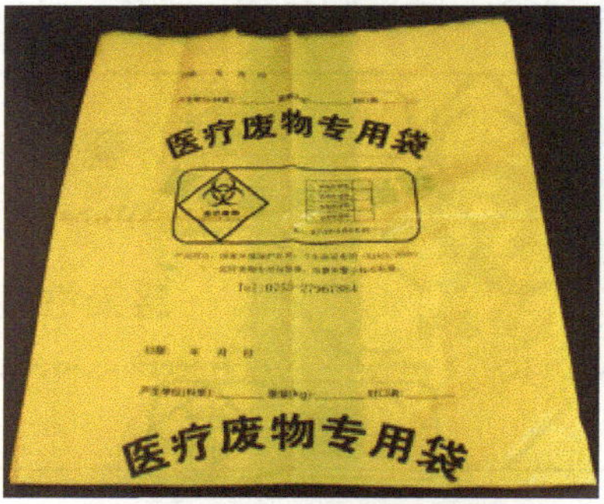

图 3–1　医疗废物包装袋

3.2 医疗废物标签应包括哪些内容?

盛装医疗废物的每个包装物、容器外表面应当有警示标志，在每个包装物、容器上应当系中文标签，中文标签的内容应包括：医疗废物产生单位、产生日期、类别及需要的特别说明等（图 3-2）。

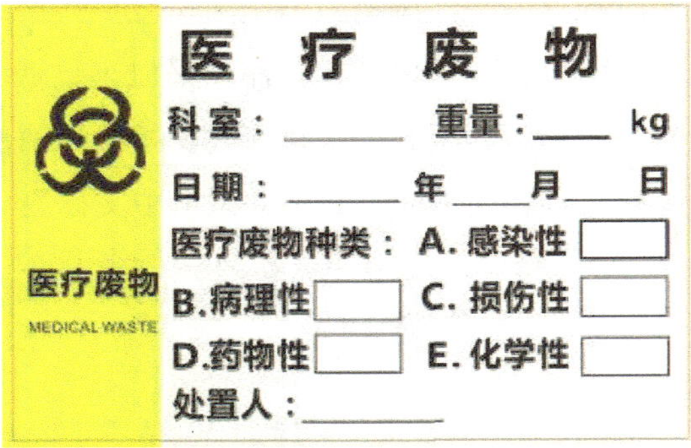

图 3-2 医疗废物标签

3.3 什么是医疗废物警示标志?

警示标志的形式为直角菱形，警告语应与警示标志组合使用。医疗废物包装袋的明显处、利器盒侧面明显处、周转箱（桶）箱体侧面或桶身明显处应印（喷）制如图 3-3 所示的警示标志和警告语。

图 3-3 医疗废物警示标志

3.4　医疗废物应该装满吗？盛装到多少时应该封口？

盛装的医疗废物达到包装物或者容器的 3/4 时，应当使用有效的封口方式，使包装物或者容器的封口紧实、严密。

3.5　放入包装物或者容器内的医疗废物可以再次取出吗？

放入包装物或者容器内的医疗废物，在收集、运送、贮存、处置的全过程均不得打开包装袋再次取出。

3.6　医疗废物包装被污染时应如何处理？

包装物或者容器的外表面被感染性废物污染时，应当对被污染处进行消毒处理或者增加一层包装。

3.7　哪些医疗废物包装应使用利器盒？

包括医用针头、缝合针、各类医用锐器、载玻片、玻璃试管、玻璃安瓿等在内的损伤性医疗废物应盛装在一次性专用硬质容器即利器盒内（图 3-4）。

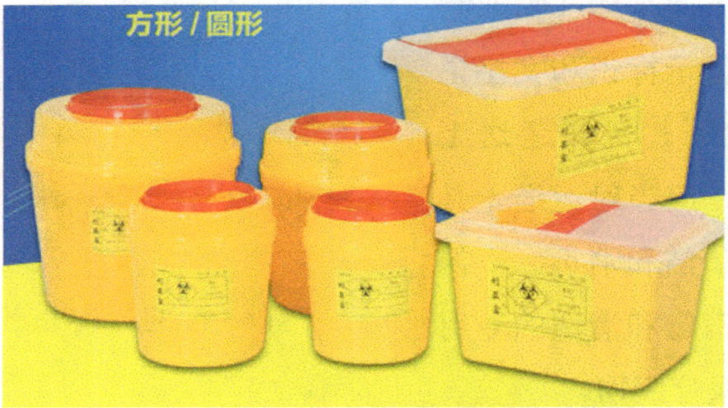

图 3-4　医疗废物利器盒

3.8 医疗废物包装容器的材料是否有特殊要求?

采用高温热处置技术处置的医疗废物，其包装袋、利器盒不应使用聚氯乙烯材料。

4 医疗卫生机构内部运送

4.1 医疗废物运送的时间、路线有哪些要求?

医疗废物应当转运到指定的暂时贮存场所。转运医疗废物的时间和路线应当相对固定，不可随意更改。转运路线应当以人流、物流最少或较偏僻为原则。转运时间应当避开诊疗高峰时段。转运过程中转运者不得离开转运车。

4.2 医疗废物运送工具必须是专用的吗?

运送医疗废物应当使用防渗漏、防遗撒、无锐利边角、易于装卸和清洁的专用运送工具（包括运送车和盛器），外表面须印（喷）制医疗废物警示标识和文字说明（图 4-1）。每天运送工作结束后，应当对运送工具及时进行清洁和消毒。不得使用未经消毒和清洗的专用工具转运医疗废物。

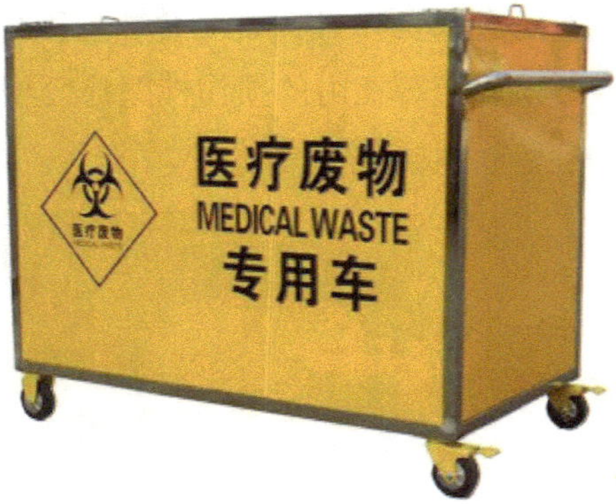

图 4-1 医疗废物运送专用车

4.3 运送医疗废物前应当检查哪些内容？

运送人员在运送医疗废物前，应当检查医疗废物分类收集是否使用专用包装袋、利器盒，每个医疗废物包装袋、利器盒上的标签内容是否符合要求，包装袋有无破损、封口是否严密。如发现不符合要求的，应当向相关部门医疗废物管理人员提出改正要求，经改正符合要求后方可运送。不得将不符合要求的医疗废物运送至暂时贮存地点。

4.4 运送医疗废物的相关规定有哪些？

《医疗废物管理条例》第十四条规定，禁止在运送过程中丢弃医疗废物。《医疗卫生机构医疗废物管理办法》第十八条规定，运送人员在运送医疗废物时，应当防止造成包装物或容器破损和医疗废物的流失、泄漏、扩散，并防止医疗废物直接接触身体。《上海市医疗废物卫生管理规范》还规定，不得在非暂时贮存场所堆放或存放医疗废物；不得将医疗废物自行运出本单位；不得接受非本医疗卫生机构产生的医疗废物。

4.5 医疗废物登记内容是什么？登记资料保存多久？

医疗卫生机构和医疗废物集中处置单位应分别对医疗废物进行登记，登记内容应当包括医疗废物的来源、种类、重量 / 数量、交接时间、处置方法、最终去向以及经办人签名等项目。登记资料至少保存 3 年，以备当地环保部门和卫生部门检查。

5 医疗废物暂时贮存

5.1 医疗废物应存放在什么场所？

《医疗废物管理条例》规定，医疗卫生机构应当建立医疗废物的暂时贮存设施、设备，不得露天存放医疗废物。

5.2 医疗废物暂时贮存设施、设备有哪些要求？

医疗卫生机构建立的医疗废物暂时贮存设施、设备应当达到以下要求：①远离医疗区、食品加工区、人员活动区和生活垃圾存放场所，方便医疗废物运送人员及运送工具、车辆的出入；②有严密的封闭措施，设专（兼）职人员管理，防止非工作人员接触医疗废物；③ 有防鼠、防蚊蝇、防蟑螂的安全措施；④防止渗漏和雨水冲刷；⑤易于清洁和消毒；⑥避免阳光直射；⑦设有明显的医疗废物警示标识和“禁止吸烟饮食”的警示标识（图 5-1）。

图 5-1 医疗废物暂存处警示标识

5.3　难以设置独立医疗废物暂时贮存库房的，医疗废物应如何存放？

不设住院病床的医疗卫生机构，如门诊部、诊所、医疗教学及科研机构，当难以设置独立的医疗废物暂时贮存库房时，应设立专门的医疗废物专用暂时贮存柜（箱），并应满足下述要求：医疗废物暂时贮存柜（箱）必须与生活垃圾存放地分开，并有防雨淋、防扬散措施，同时符合消防安全要求；将分类包装的医疗废物盛放在周转箱内后，置于专用暂时贮存柜（箱）中。柜（箱）应密闭并采取安全措施，如加锁和固定装置，做到无关人员不可移动，外部应设置警示标识。可用冷藏柜（箱）作为医疗废物专用暂时贮存柜（箱）；也可用金属或硬制塑料制作，具有一定的强度，防渗漏。

5.4　医疗废物暂时贮存场所标牌样式有何规定？

《医疗废物集中处置技术规范（试行）》中规定，应按《环境保护图形标志　固体废物贮存（处置）场》（GB 15562.2—1995）和卫生健康、生态环境主管部门制定的专用医疗废物警示标志要求，在库房外的明显处同时设置危险废物和医疗废物的警示标识（图 5-2）。

医疗废物专用警示标志	危险废物警示标志

图 5-2　危险废物和医疗废物警示标志

5.5 医疗废物在医疗卫生机构暂时贮存的时限是多久？

医疗废物应尽量做到日产日清。确实不能做到日产日清，且当地最高气温高于25℃时，应将医疗废物低温（低于20℃）暂时贮存，且时间最长不超过48h。

5.6 医疗废物暂时贮存库什么时候进行清洁和消毒处理？

医疗废物转交出去后，应当对暂时贮存地点、设施及时进行清洁和消毒处理。

医疗废物暂时贮存库房每天应在废物清运之后消毒冲洗，冲洗液应排入医疗卫生机构内的医疗废水消毒、处理系统。医疗废物暂时贮存柜（箱）应每天消毒1次。

5.7 化学性医疗废物的暂时贮存有什么特殊要求？

化学性医疗废物的暂时贮存，除符合医疗废物暂时贮存要求外，还应当符合危险化学品贮存安全要求。

5.8 病理性废物的暂时贮存有什么特殊要求？

病理性废物的暂时贮存，除符合医疗废物暂时贮存要求外，还应当具备低温贮存或者防腐条件。

6 医疗废物转移

6.1 医疗卫生机构内部医疗废物应如何交接？

在医疗卫生机构内部，医疗废物依次由本岗位医务人员、各部门分类收集管理人员、本机构医疗废物转运专职人员、本机构医疗废物暂时贮存管理人员进行交接，每个环节均应做好交接记录。最终交接给集中处置单位的收集人员。

6.2 医疗卫生机构与集中处置单位交接应执行什么管理制度？

医疗卫生机构向集中处置单位交接医疗废物应执行危险废物转移联单、医疗废物运送登记卡等管理制度。

6.3 医疗废物转移联单如何运行？

医疗卫生机构向集中处置单位转移医疗废物时，应当填写《危险废物转移联单（医疗废物专用）》。经检查、核验，接收的医疗废物包装、标识符合规定且种类、重量与转移联单所载事项相符的，医疗卫生机构的交接人员和集中处置单位的运输人员应分别在联单上签字确认。上海市实行一式四联、每月 1 份。月底，医疗卫生机构留存 1 份联单，报送给区生态环境局 1 份联单，交给集中处置单位 2 份联单；集中处置单位收到的 2 份联单，1 份留存，1 份报送给市级生态环境局。

6.4 医疗废物运送登记卡如何填写和运行？

《医疗废物运送登记卡》内容包括：运送车辆编号，运送车辆负责人，医疗卫生机构名称，感染性废物的体积、重量，损伤性废物的体积、重量，药物性废物的体积、重量，医疗卫生机构交接人员签名，处置厂医疗废物接收人员签名。《医疗废物运送登记卡》实行一车一卡，由医疗卫生机构医疗废物管理人员在交接时填写并签字。当医疗废物运至处置单位时，集中处置单位接收人员确认该登记卡上填写的医疗废物数量真实、准确后签收。

6.5 医疗废物转移运送单位应具有什么资质?

转移运送医疗废物必须由取得危险废物经营许可证的集中处置企业实施。《国家危险废物名录》对运送小型医疗卫生机构医疗废物的资质进行了豁免：从事床位总数在19张以下（含19张）的医疗卫生机构产生的医疗废物的收集活动，收集过程不按危险废物管理。

6.6 医疗废物运送车辆应符合什么要求?

医疗废物运送车辆应当遵守国家危险货物运输管理的规定，符合《医疗废物转运车技术要求（试行）》（GB 19217—2003），有明显医疗废物标识，达到防渗漏、防遗撒以及其他环境保护和卫生要求。车辆使用后，应当在医疗废物集中处置场所内及时进行消毒和清洁。运送医疗废物的专用车辆不得运送其他物品（图6-1）。

图6-1 医疗废物转运车

6.7 医疗废物运送车辆应随车配备什么？

运送车辆应配备：①《医疗废物集中处置技术规范（试行）》规范文本；②《危险废物转移联单》（医疗废物专用）；③《医疗废物运送登记卡》；④运送路线图；⑤通信设备；⑥医疗废物产生单位及其管理人员名单与电话号码；⑦事故应急预案及联络单位和人员的名单、电话号码；⑧收集医疗废物的工具、消毒器具与药品；⑨备用的医疗废物专用袋和利器盒；⑩备用的人员防护用品。

6.8 医疗废物运送车辆清洗消毒有何规定？

医疗废物运送专用车每次运送完毕，应在处置单位内对车厢内壁进行消毒，喷洒消毒液后密封至少 30 min。医疗废物运送的周转箱每次运送完毕，应在医疗卫生机构或医疗废物处置单位内对周转箱进行消毒、清洗。车辆清洗晾干后方可再次投入使用。

6.9 医疗废物可以通过水路、铁路、航空运输吗？

《医疗废物管理条例》第十五条规定，禁止邮寄医疗废物。禁止通过铁路、航空运输医疗废物。有陆路通道的，禁止通过水路运输医疗废物；没有陆路通道必须经水路运输医疗废物的，应当经设区的市级以上人民政府环境保护行政主管部门批准，并采取严格的环境保护措施后，方可通过水路运输。禁止在饮用水水源保护区的水体上运输医疗废物。

7 医疗废物处置

7.1 医疗废物集中处置单位应取得何种资质?

凡从事医疗废物集中处置活动的单位，必须取得危险废物经营许可证，未取得许可证的单位不得从事相关活动。

7.2 集中处置单位贮存医疗废物的时限有多久?

进入处置单位的医疗废物若不能立即处置，应盛装于周转箱内贮存于医疗废物暂时贮存库房中。当处置单位医疗废物暂时贮存温度≥5℃，医疗废物暂时贮存时间不得超过24h；当医疗废物暂时贮存温度<5℃，医疗废物暂时贮存时间不得超过72h。

7.3 医疗废物有哪些处置方式?

医疗废物处置方式有高温焚烧、高温蒸汽消毒、微波消毒、化学消毒等。

7.4 医疗废物焚烧残余物是否属于危险废物?

医疗废物焚烧残余物主要有除尘设备产生的飞灰、焚烧炉的炉渣。飞灰属于危险废物，炉渣不属于危险废物。

7.5 医疗废物处置单位应在哪些位置、设置何种警示标牌?

医疗废物处置单位应在厂区出入口、暂时贮存设施、处置场所等，同时设置危险废物警示标志和医疗废物警示标志。

7.6 农村地区的医疗废物如何管理和处置？

根据《医疗废物管理条例》第二十一条，不具备集中处置医疗废物条件的农村，医疗卫生机构应当按照县级人民政府卫生行政主管部门和环境保护行政主管部门的要求，自行就地处置其产生的医疗废物。自行处置医疗废物的，应当符合下列基本要求：使用后的一次性医疗器具和容易致人损伤的医疗废物，应当消毒并作毁形处理；能够焚烧的，应当及时焚烧；不能焚烧的，消毒后集中填埋。

8 医疗废物处理过程中的 EHS 管理

8.1 EHS 管理包括哪些方面？

EHS 是环境（Environment）、健康（Health）、安全（Safety）的缩写。EHS 管理体系是环境管理体系（EMS）和职业健康安全管理体系（OHSMS）两体系的整合。

8.2 医疗废物处置应符合哪些污染控制标准？

医疗废物处理过程应严格控制对环境可能造成危害的水、气、声、渣等环境因素，各类污染物的排放标准必须符合国家、地方和行业有关废水、废气、噪声、固体污染物控制标准，如《大气污染物综合排放标准》（GB 16297—1996）、《危险废物焚烧污染控制标准》（GB 18484—2001）、《恶臭污染物排放标准》（GB 14554—93）、《污水综合排放标准》（GB 8978—1996）、《污水排入城镇下水道水质标准》（GB/T 31962—2015）、《工业企业厂界环境噪声排放标准》（GB 12348—2008）等。

8.3 医疗废物集中处置单位是否需要安装污染物排放自动监测系统？

高温焚烧医疗废物的集中处置单位，应安装固定污染源烟气排放连续自动监测系统。监控项目至少包含二氧化硫、氮氧化物、颗粒物、一氧化碳、含氧量、氯化氢、烟气温度、烟气压力、烟气流速或流量、烟气含湿量等。自动监测数据可用于环境执法。

8.4 医疗废物处置从业人员职业卫生防护的总体要求是什么？

《医疗废物管理条例》第十条规定，医疗卫生机构和医疗废物集中处置单位，应当采取有效的职业卫生防护措施，为从事医疗废物收集、运送、贮存、处置等工作的人员和管理人员配备必要的防护用品，定期进行健康检查；必要时，对有关人员进行免疫接种，防止其受到健康损害。

8.5 医疗废物收运人员如何做好职业卫生防护？

收运人员在运送医疗废物过程中须穿戴防护手套、口罩、工作服、靴等防护用品（图 8-1）；收运人员体检频次：2 次 / 年，必要时进行预防性免疫接种。

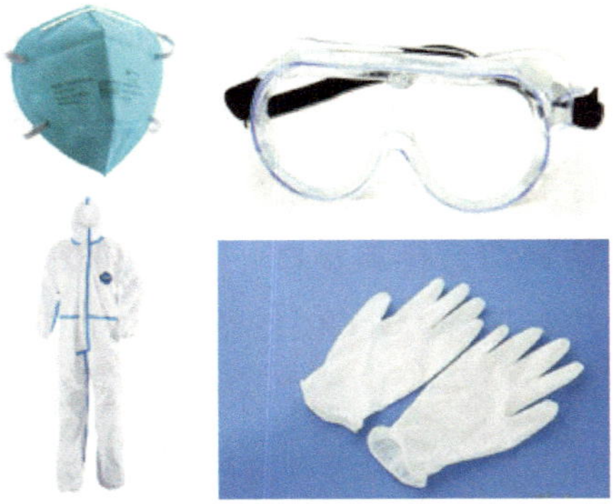

图 8-1　常用个人防护用品

8.6 医疗废物处置单位操作人员如何做好职业卫生防护？

《医疗废物集中处置技术规范（试行）》要求：操作人员在操作过程中须穿戴防护手套、口罩、工作服、靴等防护用品，如有液体或熔融物溅出危险时，还需配戴护目镜。

8.7 涉及医疗废物的仪器设备和防护用品如何进行消毒？

仪器设备消毒：使用 0.2% 过氧乙酸、500 mg/L 二氧化氯溶液，或有效氯（溴）的消毒溶液擦拭消毒。防护用品消毒：耐热的用品可使用流通蒸汽消毒 20 ～ 30 min 或用压力蒸汽 121 ℃作用 20 ～ 30 min，耐湿的用品（包括防护眼镜）使用有效氯含量为 1 000 mg/L 的消毒液浸泡 30 min。

8.8 医疗废物运送工具使用什么消毒方法？

运送车辆、周转箱（桶）、贮存设施及其他有关的物品表面、环境消毒：使用 0.2% 过氧乙酸溶液或有效氯含量为 1 000 ～ 2 000 mg/L 的消毒液进行喷洒、喷雾或擦拭，作用 60 min 后洗净。地面消毒：使用 500 ～ 1 000 mg/L 的二氧化氯溶液，或有效氯（溴）含量为 1 000 ～ 2 000 mg/L 的消毒液进行喷洒或拖地（图 8-2）。

图 8-2 医疗废物转运车辆消毒

9 医疗废物监督管理

9.1 医疗废物全过程管理涉及的部门有哪些?

医疗废物全过程管理涉及地方人民政府、卫生健康主管部门、生态环境主管部门、医疗卫生机构、医疗废物集中处置单位、交通运输部门等。

9.2 医疗废物管理中省级人民政府的工作职责是什么?

省级人民政府负责做好本省医疗废物集中处置设施建设规划。

9.3 医疗废物管理中卫生健康主管部门的工作职责是什么?

卫生健康主管部门负责医疗卫生机构管理，对医疗废物收集、运送、贮存、处置活动中的疾病防治工作实施统一监督管理。

9.4 医疗废物管理中生态环境主管部门的工作职责是什么?

生态环境主管部门负责集中处置单位管理，对医疗废物收集、运送、贮存、处置活动中的环境污染防治工作实施统一监督管理。

9.5 医疗废物管理中交通运输主管部门的工作职责是什么?

交通运输主管部门负责医疗废物转移运送过程中的交通运输管理。

9.6 医疗废物管理中医疗卫生机构的工作职责是什么？

医疗卫生机构法定代表人是医疗卫生机构废弃物分类和管理的第一责任人，产生废弃物的具体科室和操作人员是直接责任人。医疗卫生机构的工作职责具体包括医疗卫生机构医疗废物管理责任制的建立，监控部门和专（兼）职管理人员的配备，管理制度和应急方案的落实，工作人员职业卫生安全防护和培训的开展，医疗废物分类收集、内部运送、交接登记、暂时贮存、集中处置等工作，防止医疗废物流失、泄漏、扩散。

9.7 医疗废物集中处置单位的工作职责是什么？

医疗废物集中处置单位应保障医疗废物处置具备基本运行条件，保障医疗废物处置设施运行稳定，保障污染防治设施的配置及运行效果，以及落实安全生产和劳动保护措施等。

9.8 医疗卫生机构医疗废物监督检查如何进行？

可以以《医疗废物专项监督检查表格》形式，重点监督检查各级各类医疗卫生机构医疗废物管理责任制的建立，监控部门和专（兼）职管理人员的配备，管理制度和应急方案的落实，工作人员职业卫生安全防护和培训的开展，医疗废物分类收集、内部运送、交接登记、暂时贮存、集中处置等工作，防止医疗废物流失、泄漏、扩散。

9.9 医疗废物集中处置单位监督检查如何进行？

根据《医疗废物集中焚烧处置设施运行监督管理技术规范》（试行），生态环境主管部门可通过书面检查、现场核查以及远程监控等方式实施对医疗废物集中焚烧处置设施运行的监督管理。监督管理包括准备、检查、监测、综合分析、意见反馈、整改和复查七个阶段。可以以《危险废物（含医疗废物）经营单位规范化管理指标及抽查表》的形式，逐项检查危险废物识别标志设置情况，危险废物管理计划制定情况，危险废物申报登记、转移联单、经营许可、应急预案备案等管理制度执行情况，贮存、利用、处置危险废物是否符合相关标准规范等。

9.10 医疗卫生机构违反医疗废物管理规定有哪些法律责任?

①被责令停止违法行为，给予警告，限期改正；②处以不同数额的罚款；③没收违法所得；④暂扣或者吊销执业许可证件；⑤承担刑事责任。

9.11 集中处置单位违反医疗废物管理规定有哪些法律责任?

①被责令停止违法行为，给予警告，限期改正；②处以不同数额的罚款；③没收违法所得；④暂扣或者吊销危险废物经营许可证；⑤承担刑事责任。

Part 3

第三部分

疫情期间医疗废物管理

10 疫情期间医疗废物管理规定

10.1 什么是新型冠状病毒感染的肺炎疫情时期涉疫医疗废物？

《国家卫生健康委办公厅关于做好新型冠状病毒感染的肺炎疫情期间医疗卫生机构医疗废物管理工作的通知》（国卫办医函〔2020〕81号）对此次新冠疫情医疗废物进行了明确：医疗卫生机构在诊疗新型冠状病毒感染的肺炎患者及疑似患者发热门诊和病区（房）产生的废弃物，包括医疗废物和生活垃圾，均应当按照医疗废物进行管理。

10.2 疫情期间国家出台了哪些关于医疗废物的文件？

国家卫生健康委办公厅发布了《国家卫生健康委办公厅关于做好新型冠状病毒感染的肺炎疫情期间医疗卫生机构医疗废物管理工作的通知》，生态环境部发布了《新型冠状病毒感染的肺炎疫情医疗废物应急处置管理与技术指南（试行）》，部分地方政府也出台了一些医疗废物管理政策。国家卫生健康委要求，要加强对提供后勤服务的机构和人员的管理，组织开展培训，督促其掌握医疗废物管理的基本要求，切实履行职责，及时处理产生的医疗废物。

10.3 疫情期间医疗废物处置能力不足的地区应如何处置？

《应对甲型H1N1流感疫情医疗废物管理预案》提出，医疗废物处置能力不足的地区，应按照应急方案处置医疗废物；因特殊原因确实不具备集中处置医疗废物条件的地区，特别是农村偏远地区，医疗卫生机构可对医疗废物进行就地焚烧处置。

《新型冠状病毒感染的肺炎疫情医疗废物应急处置管理与技术指南（试行）》提出，对存在医疗废物处置能力缺口的地市，协调其他地市建立应急处置跨区域协同机制。因特殊原因，不具备集中处置条件的，可根据当地人民政府确定的方案对医疗废物进行就地焚烧处置。

10.4　为便利应急处置，疫情医疗废物可做何种豁免？

《新型冠状病毒感染的肺炎疫情医疗废物应急处置管理与技术指南（试行）》提出，医疗卫生机构自行或在邻近医疗卫生机构采用可移动式医疗废物处置设施应急处置医疗废物，可豁免环境影响评价、医疗废物经营许可等手续，但应合理设置处置地点，避让饮用水水源保护区、集中居住区等环境敏感区，并在设区的市级卫生健康主管部门和生态环境主管部门报备。可移动式医疗废物处置设施供应商应确保医疗废物处置效果满足相关标准和技术规范要求。

10.5　疫情期间如何利用媒体做好医疗废物管理宣传？

首先，注重宣传引导，动态把握疫情期间医疗废物管理与处置舆情方向。根据《新型冠状病毒感染的肺炎疫情医疗废物应急处置管理与技术指南（试行）》要求，地方各级生态环境主管部门应根据本级人民政府的有关要求做好相关信息发布工作。辖区内生态环境部门应积极与主流媒体及时做好沟通，加强宣传，及时释放信息，定期报送疫情防控相关环保工作进展和成效，大力宣传疫情防控先进典型，弘扬正能量。其次，充分利用政府、生态环境部门官方网站、微博、微信公众号等多个媒介平台发布辖区内医疗废物监管动态，及时让公众掌握信息；积极向有影响力的新闻媒体发声，及时向公众宣传医疗废物管理相关知识，介绍开展疫情防控的系列工作，普及疫情防控环保知识。最后，利用大数据平台，收集掌握医疗废物管理、收集处置的新闻舆情，及时发现负面新闻，迅速做出应对措施，解决公众和媒体反映的突出问题。

10.6 疫情期间医疗卫生机构医疗废物管理主体责任如何落实？

医疗卫生机构要高度重视新型冠状病毒感染的肺炎疫情期间医疗废物管理，切实落实主体责任，其法定代表人是医疗废物管理的第一责任人，产生医疗废物的具体科室和操作人员是直接责任人。实行后勤服务社会化的医疗卫生机构要加强对提供后勤服务的机构和人员进行管理，组织开展培训，督促其掌握医疗废物管理的基本要求，切实履行职责。加大环境卫生整治力度，及时处理产生的医疗废物，避免各种废弃物堆积，努力创造健康卫生环境。

11 疫情期间医疗废物收集暂存

11.1 疫情期间医疗卫生机构如何加强医疗废物的分类收集？

（1）明确分类收集范围。医疗卫生机构在诊疗新型冠状病毒感染的肺炎患者及疑似患者发热门诊和病区（房）产生的废弃物，包括医疗废物和生活垃圾，均应当按照医疗废物进行分类收集。

（2）规范包装容器。医疗废物专用包装袋、利器盒的外表面应当有警示标识，在盛装医疗废物前，应当进行认真检查，确保其无破损、无渗漏。医疗废物收集桶应为脚踏式并带盖。医疗废物装到包装袋或者利器盒的 3/4 时，应当有效封口，确保封口严密。应当使用双层包装袋盛装医疗废物，采用鹅颈结式封口，分层封扎。

（3）做好安全收集。按照医疗废物类别及时分类收集，确保人员安全，控制感染风险。盛装医疗废物的包装袋和利器盒的外表面被感染性废物污染时，应当增加一层包装袋。分类收集使用后的一次性隔离衣、防护服等物品时，严禁挤压。每个包装袋、利器盒应当系有或粘贴中文标签，标签内容包括医疗废物产生单位、产生部门、产生日期、类别等，并在特别说明中标注“新型冠状病毒感染的肺炎”或者简写为“新冠”。

（4）分区域进行处理。收治新型冠状病毒感染的肺炎患者及疑似患者发热门诊和病区（房）的潜在污染区和污染区产生的医疗废物，在离开污染区前应当对包装袋表面采用 1 000 mg/L 的含氯消毒液进行喷洒消毒（注意喷洒均匀）或在其外面加套一层医疗废物包装袋；清洁区产生的医疗废物按照常规的医疗废物处置。

（5）做好病原标本处理。医疗废物中含病原体的标本和相关保存液等高危险废物，应当在产生地点进行压力蒸汽灭菌或者化学消毒处理，然后按照感染性废物收集处理。

11.2 疫情期间医疗废物分类收集有哪些特殊要求？

《国家卫生健康委办公厅关于做好新型冠状病毒感染的肺炎疫

情期间医疗卫生机构医疗废物管理工作的通知》对此次新冠肺炎疫情医疗废物收集做了进一步要求：医疗废物收集桶应为脚踏式并带盖。分类收集使用后的一次性隔离衣、防护服等物品时，严禁挤压。

《新型冠状病毒感染的肺炎疫情医疗废物应急处置管理与技术指南（试行）》明确提出，对新冠肺炎疫情防治过程中产生的感染性医疗废物进行消毒处理，严格按照《医疗废物专用包装袋、容器和警示标志标准》进行包装，再置于指定周转桶（箱）或一次性专用包装容器中（图 11-1）。

图 11-1　疫情期间医疗废物收集

11.3　隔离观察点废弃的个人防护用品属于医疗废物吗?

《上海市新型冠状病毒肺炎防控方案（第四版）》中提出：隔离观察（留验）点工作人员和服务人员脱卸的个人防护用品应放入医疗废物垃圾袋内，扎紧作为医疗废物处置。

11.4 隔离观察点产生的生活垃圾属于医疗废物吗?

《上海市新型冠状病毒肺炎防控方案（第四版）》中提出：无相关症状的观察隔离人员所有产生的生活垃圾，收集在市容环卫部门提供的专用垃圾袋中，使用 2 000 mg/L 有效氯（溴）消毒溶液喷洒后，作为一般生活垃圾由环卫部门专人收集，专车运送到指定的垃圾焚烧厂焚烧；有发热等相关症状的观察隔离人员所有产生的生活垃圾，收集在感染性的医用废物垃圾袋内，使用双层包装袋，用 10 000 mg/L 有效氯（溴）含氯（溴）消毒溶液喷洒后，采用鹅颈结式封口，分层封扎，由区安排专车送至附近设有发热门诊的医疗卫生机构按感染性医疗废物处置。

11.5 疫情期间使用过的口罩属于医疗废物吗?

根据国家卫生健康委印发的《新型冠状病毒感染不同风险人群防护指南》，健康人群使用后的口罩按照生活垃圾分类的要求处理；疑似病例或确诊患者佩戴的口罩，不可随意丢弃，应视为医疗废物，严格按照医疗废物有关流程处理。

11.6 疫情期间首诊隔离点生活垃圾属于医疗废物吗?

《国家卫生健康委办公厅关于加强新冠肺炎首诊隔离点医疗管理工作的通知》（国卫办医函〔2020〕120 号）要求，首诊隔离点是地方政府指定的，在医疗卫生机构以外用于收治新冠肺炎疑似病例轻症患者的，具备一定条件的宾馆、酒店、招待所等场所。首诊隔离点在诊疗活动中产生的废弃物，包括医疗废物和生活垃圾，均应当按照医疗废物，参照《国家卫生健康委办公厅关于做好新型冠状病毒感染的肺炎疫情期间医疗卫生机构医疗废物管理工作的通知》进行管理。

11.7 疫情期间医疗废物包装有哪些特殊要求?

《国家卫生健康委办公厅关于做好新型冠状病毒感染的肺炎疫情期间医疗卫生机构医疗废物管理工作的通知》对此次新冠肺炎疫情医疗废物包装进行了进一步要求：应当使用双层包装袋盛装医疗废物，采用鹅颈结式封口，分层封扎。盛装医疗废物的包装袋和利器盒的外表面被感染性废物污染时，应当增加一层包装袋。每个包装袋、利器盒标签的特别说明中应标注“新型冠状病毒感染的肺炎”或者简写为“新冠”。收治新型冠状病毒感染的肺炎患者及疑似患者发热门诊和病区（房）的潜在污染区和污染区产生的医疗废物，在离开污染区前应当对包装袋表面采用 1 000 mg/L 的含氯消毒液喷洒消毒（注意喷洒均匀）或在其外面加套一层医疗废物包装袋。

11.8 疫情期间医疗卫生机构如何加强医疗废物的运送贮存?

（1）安全运送管理。在运送医疗废物前，应当检查包装袋或者利器盒的标识、标签以及封口是否符合要求。工作人员在运送医疗废物时，应当防止造成医疗废物专用包装袋和利器盒的破损，防止医疗废物直接接触身体，避免医疗废物泄漏和扩散。每天运送结束后，对运送工具进行清洁和消毒，含氯消毒液浓度为 1 000 mg/L；运送工具被感染性医疗废物污染时，应当及时消毒处理。

（2）规范贮存交接。医疗废物暂存处应当有严密的封闭措施，设有工作人员进行管理，防止非工作人员接触医疗废物。医疗废物宜在暂存处单独设置区域存放，尽快交由医疗废物处置单位进行处置。用 1 000 mg/L 的含氯消毒液对医疗废物暂存处地面进行消毒，每天两次。医疗废物产生部门、运送人员、暂存处工作人员以及医疗废物处置单位转运人员之间，要逐层登记交接，并说明其来源于新型冠状病毒感染的肺炎患者或疑似患者。

（3）做好转移登记。严格执行危险废物转移联单管理，对医疗废物进行登记。登记内容包括医疗废物的来源、种类、重量/数量、交接时间、最终去向以及经办人签名，特别注明“新型冠状病毒感染的肺炎”或“新冠肺炎”，登记资料保存 3 年。

医疗卫生机构要及时通知医疗废物处置单位进行上门收取，并做好相应记录。各级卫生健康主管部门和医疗卫生机构要加强与生态环境主管部门、医疗废物处置单位的信息互通，配合做好新型冠状病毒感染的肺炎疫情期间医疗废物的规范处置。

11.9 疫情期间医疗废物暂时贮存场所有什么特殊要求?

《新型冠状病毒感染的肺炎疫情医疗废物应急处置管理与技术指南（试行）》要求，有条件的医疗卫生机构可对肺炎疫情防治过程产生的感染性医疗废物的暂时贮存场所实行专场存放、专人管理，不与其他医疗废物和生活垃圾混放、混装。

11.10 疫情期间医疗废物暂时贮存时限有什么特殊要求?

《新型冠状病毒感染的肺炎疫情医疗废物应急处置管理与技术

指南（试行）》要求，新冠疫情期间医疗废物在医疗卫生机构内暂存时间不超过 24h。

11.11　疫情期间医疗废物暂时贮存场所消毒有什么要求？

医疗卫生机构医疗废物暂时贮存场所由专人使用 0.2% ～ 0.5% 过氧乙酸或 1 000 ～ 2 000 mg/L 含氯消毒剂喷洒墙壁或拖地消毒，每天上午、下午各一次。

11.12　疫情期间医疗废物登记有哪些特殊要求？

《国家卫生健康委办公厅关于做好新型冠状病毒感染的肺炎疫情期间医疗卫生机构医疗废物管理工作的通知》要求，医疗废物产生部门、运送人员、暂存处工作人员以及医疗废物处置单位转运人员之间，要逐层登记交接，并说明其来源于新型冠状病毒感染的肺炎患者或疑似患者。

《新型冠状病毒感染的肺炎疫情医疗废物应急处置管理与技术指南（试行）》要求，有条件的地区，可安排固定专用车辆单独运输新冠肺炎疫情防治过程产生的感染性医疗废物，并与其他医疗废物分开填写转移联单，建立台账。

11.13　疫情期间临时性医院建设暂时贮存仓库有哪些要求？

应按照传染病医院的医疗废物暂时贮存仓库要求进行建设，不能因临时性而降低建设标准。主要要求有：远离医疗区、食品加工区、人员活动区和生活垃圾存放场所，方便医疗废物运送人员及运送工具、车辆的出入，有严密的封闭措施，设专（兼）职人员管理，防止非工作人员接触医疗废物，有防鼠、防蚊蝇、防蟑螂的安全措施，防止渗漏和雨水冲刷，易于清洁和消毒，避免阳光直射，设有明显的危险废物警示标识、医疗废物警示标识和“禁止吸烟饮食”的警示标识。暂时贮存病理性废物，应当具备低温贮存或者防腐条件。

12 疫情期间医疗废物处置

12.1 新冠肺炎患者产生的医疗废物处置方式应如何选择?

新型冠状病毒感染的肺炎患者产生的医疗废物，宜采用高温焚烧方式处置。

12.2 疫情期间医疗废物应急处置技术路线应如何确定?

将新冠肺炎疫情防治过程中产生的医疗废物与其他医疗废物实行分类、分流管理。医疗废物集中处置单位应优先处置疫情医疗废物，可将其他医疗废物转移至应急处置设施进行处置；也可以按照应急处置跨区域协同机制，将新冠肺炎医疗废物转运至邻近地区医疗废物集中处置设施进行处置。在确保处置效果的前提下，可选取可移动式医疗废物专业处置设施和非医疗废物专业处置设施作为医疗废物应急处置的备选资源。可移动式医疗废物专业处置包括焚烧、高温蒸汽消毒、微波消毒、化学消毒等方式。非医疗废物专业处置设施是指采用高温焚烧技术的固体废物处置设施，如危险废物焚烧设施（焚烧温度高于 1 100℃）、生活垃圾焚烧设施（焚烧温度高于 850℃）、炉窑温度高于 850℃的工业炉窑等。

因特殊原因确实不具备集中处置医疗废物条件的地区，特别是农村偏远地区，医疗卫生机构可对医疗废物进行就地焚烧处置。

12.3 非专业处置设施应急处置医疗废物有哪些技术要求和注意事项？

使用危险废物焚烧设施、生活垃圾焚烧设施、工业炉窑等非医疗废物专业处置设施开展新冠肺炎疫情医疗废物应急处置活动，一是需要按照卫生健康主管部门的要求切实做好卫生防疫工作；二是应针对医疗废物划定专门卸料接收区域、清洗消毒区域，增加必要的防雨防淋、防泄漏措施，对医疗废物运输车辆规划专用行车路线，并配置专人管理；三是医疗废物接收现场应设置警示、警戒限制措施；四是进料方式宜采用专门输送上料设备，防止医疗废物与其他焚烧物接触造成二次交叉污染；五是注意做好医疗废物与其他焚烧物的进料配伍，保持工艺设备运行平稳可控；六是技术操作人员应接受必要的技术培训。

12.4 应急处置医疗废物的生活垃圾焚烧设施如何选定？炉型如何选择？

《生活垃圾焚烧设施应急处置肺炎疫情医疗废物工作相关问题及解答》中指出，应急处置医疗废物的生活垃圾焚烧设施，由县级以上地方人民政府组织卫生健康、生态环境、住房与城乡建设等部门共同研究选定。

采用生活垃圾焚烧设施应急处置医疗废物的，可以采用炉排式焚烧炉，不宜采用流化床式焚烧炉。医疗废物与生活垃圾混合焚烧每 h 比例控制在 5% 以内。

12.5 疫情期间医疗废物收运有哪些特殊要求?

《医疗废物集中处置技术规范（试行）》要求处置单位在运送医疗废物时必须使用固定专用车辆。

《新型冠状病毒感染的肺炎疫情医疗废物应急处置管理与技术指南（试行）》指出，肺炎疫情防治过程产生的感染性医疗废物的运输使用专用医疗废物运输车辆，或使用参照医疗废物运输车辆要求进行临时改装的车辆；规划专用行车路线，运输路线尽量避开人口稠密地区，运输时间避开上下班高峰期（图 12-1）。

图 12-1 上海市医疗废物应急收运

12.6 疫情期间医疗废物对于集中处置单位贮存时限有什么特殊要求?

《新型冠状病毒感染的肺炎疫情医疗废物应急处置管理与技术

指南（试行）》要求，新冠肺炎疫情期间，运抵处置场所的医疗废物尽可能做到随到随处置，在处置单位的暂时贮存时间不超过 12h。

12.7　疫情期间是否需要设置医疗废物处置的隔离区?

《新型冠状病毒感染的肺炎疫情医疗废物应急处置管理与技术指南（试行）》指出，处置单位内必须设置医疗废物处置的隔离区，隔离区应有明显的标识，无关人员不得进入；处置单位隔离区必须由专人负责，按照卫生健康主管部门要求的方法和频次对墙壁、地面、物体表面进行喷洒或拖地消毒。

12.8　疫情期间医疗废物处置场所消毒有什么要求?

医疗废物处置隔离区必须由专人使用 0.2% ～ 0.5% 过氧乙酸或 1 000 ～ 2 000 mg/L 含氯消毒剂对墙壁、地面或物体表面进行喷洒或拖地消毒，每天上午、下午各一次（图 12-2）。

图 12–2　医疗废物消毒

12.9 疫情期间职业卫生防护有哪些特殊要求?

《新型冠状病毒感染的肺炎疫情医疗废物应急处置管理与技术指南（试行）》指出，疫情期间要加强操作人员的日常体温监测工作。有条件的地区，可安排收集、贮存、转运、处置医疗废物的一线操作人员集中居住。

《新型冠状病毒肺炎暴露风险防范手册——环保相关从业人员》中提出，处理 / 处置医疗废物的工作人员在运送或处置操作完成之后要立即洗手和消毒，从特殊区域返回的人员应当洗澡；离开工作区前，工作人员必须使用含氯消毒液对使用过的工具进行消毒（图 12-3）。

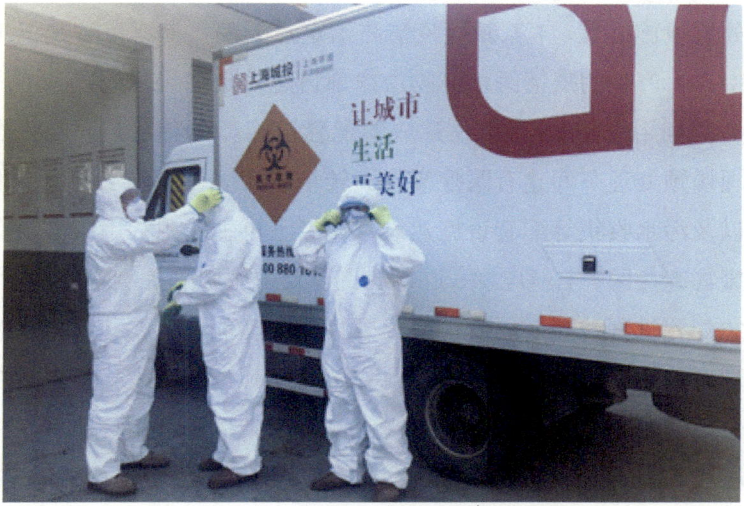

图 12-3 上海市应急收运人员个人防护

13 疫情期间应急处置案例

13.1 利用生活垃圾焚烧设施应急处置医疗废物的案例

新冠疫情期间，上海市为应急处置医疗废物，在上海市生态环境局、市卫生健康委、市绿化市容局的支持下，上海市固体废物处置有限公司按照《上海市医疗废物应急处置保障预案》利用老港再生能源利用中心生活垃圾焚烧设施，开展常规医疗废物应急处置工作，每天约40t。1家新冠疫情防治定点医院（儿科医院）和76家门诊发热医院（含相关隔离点送来的疑似病人产生的疫情医疗废物）产生的10t以内的疫情医疗废物，仍在公司本部的专门医疗废物焚烧处置设施处置，1家新冠疫情防治定点医院（上海市公共卫生临床中心）产生的疫情医疗废物由内部自建的焚烧设施处置。本次应急处置利用了2014年就已建成的配套设施，在老港一般工业固废填埋场预处理车间建有医疗废物周转箱装卸区、清洗消毒区、堆放区以及污水收集等配套设施建设，在老港再生能源利用中心建有医疗废物卸货区、上料电梯、进料装置以及卫生防疫控制等配套设施。应急处置过程按照《上海市医疗废物卫生管理规范》和《医疗废物管理有关物品消毒方法》进行卫生防疫控制，医疗废物与生活垃圾混合焚烧每小时比例控制在5%以内，污染物排放指标执行《生活垃圾焚烧大气污染物排放标准》（DB 31/768—2013）。上海市固体废物处置有限公司制定了《上海市新型冠状病毒肺炎疫情期间医疗废物收运处置应急保障方案》，考虑在本部基地3台焚烧炉全部故障停炉等极端情况下，将疫情医疗废物也送老港再生能源利用中心生活垃圾焚烧设施应急处置。疫情期间该极端情况未实际发生。

图 13-1　上海市医疗废物应急处置设施场景

图 13-2　医疗废物应急处置上料

13.2　利用水泥窑协同处置医疗废物的案例

新冠肺炎疫情暴发后，华新水泥旗下多家公司进入应急处置状态，认真做好相关设备调试和人员操作演练，随时接收和处置医疗废物。2020 年 2 月初华新水泥下属华新环境公司接到医疗废物应急处置任务。该公司及各协同处置工厂立即召回相关技术人员和操作人员，设置隔离区，有序开展防护物资调配、人员演练等准备工作，对处置过程中的消毒、收运、投递、处置等每一个环节都进行了周密部署。在医疗废物处置中，该公司组织多台全封闭、防渗漏、带 GPS 定位的集装箱卧式自卸车来承担医疗废物运输任务，每车还配备了引导车，确保运输过程万无一失。同时，在处置过程中不设中转点、不滞留，实行日清日结，既保障了员工安全和处置过程的安全、高效，也彻底消除了医疗废物二次污染风险。2020 年 2 月 4 日—3 月 6 日，累计处置医疗废物 55 t（图 13-3）。

图 13-3　华新环境公司应急处置医疗废物

13.3　利用移动式医疗废物应急处置系统的案例

案例一：启迪环境科技发展股份有限公司移动式医疗废物处理系统参与湖北省武汉市新冠肺炎疫情医疗废物应急处置工作，负责协和江南医院（武汉市江夏区第一人民医院）医疗废物应急处置，并兼顾雷神山医院医疗废物应急处置工作。该系统采用撬装式可定置或车载高温蒸汽医疗废物处置设施，主要由医疗废物灭菌器、灭菌小车、冷却循环水辅助系统、破碎机、卸料机、自动进出车机构（或气缸推进装置）、升降机等设备组成。设备自身配有蒸汽发生器和空气压缩机，无须外接汽源，只需接入动力电和水源即可，方便进入疫情区医疗卫生机构。设备处理能力为 1.5t/d（图 13-4）。

案例二：为全力支持中国抗击疫情，联合国工业发展组织和生态环境部对外合作与交流中心紧急合作，开发实施了“新型冠状病毒疫情中国应急响应项目”。由该项目支持的移动式高温蒸汽医疗废物应急处理设备在武汉市洪山区投入使用，该设备服务范围覆盖洪山区所有新冠肺炎定点医院，负责处理具有高风险、高感染性的医疗废物，处理能力可达 1.8t/d。考虑到疫区实际需求，结合此前完成的全球环境基金中国医疗废物环境可持续管理项目技术经验，采用车载移动式高温蒸汽医疗废物处理技术，既降低了医疗废物长距离运输过程中的传染风险，又有效避免了二噁英的产生和大气排放，不仅能在疫情期间提供机动现场处理服务，还可作为集中处置设备检修期间备用处理能力的补充，实现了长期提升疫区医疗废物应急处置能力的目标。

图 13-4　医疗废物应急处理系统

13.4 应急运输医疗废物的案例

为确保新冠肺炎定点治疗医院的医疗废物在收集转运和处置等环节的安全，武汉疫情防控指挥部部署消防、环保和医院多部门按照相关规范要求共同做好医疗废物收集转运处置工作。武汉市消防救援支队与45家医疗单位对接转运医疗废物（图13-5）。

图13-5 消防参与转运医疗废物

Series of Books on the Implementation of International Environmental Conventions

Research Results of Foreign Environmental Cooperation Center, Ministry of Ecology and Environment of China

FECO

Output of United Nations Industrial Development Organization Project

UNIDO UNITED NATIONS INDUSTRIAL DEVELOPMENT ORGANIZATION

Training Guidance for the Management of Medical Waste during the COVID-19 Outbreak in China

Chief Editor

Foreign Environmental Cooperation Center, Ministry of Ecology and Environment of China

National Medical Waste Management Training Base (Shanghai)

United Nations Industrial Development Organization

China Environment Publishing Group·Beijing

Acknowledgements

The compiling of the *Training Guidance for the Management of Medical Waste during the COVID-19 Outbreak in China* was financially supported by the project "Emergency Response Assistance to the Outbreak of Coronavirus Disease 2019 in China" jointly developed and implemented by the Foreign Environmental Cooperation Center (FECO) of the Ministry of Ecology and Environment of China and the United Nations Industrial Development Organization (UNIDO), with guidance from the Ministry of Commerce of China. Special thanks to Mr. Zhou Yunrui (UNIDO) and his project team.

This book was developed with guidance from the Department of International Cooperation of the Ministry of Ecology and Environment, especially Mr. Guo Jing (Director General), Ms. Zhang Jieqing and Mr. Xiao Xuezhi (Deputy Director Generals), Mr. Sun Xuefeng, Mr. Zhuo Zhuang, Ms. Chen Haijun, Mr. Zhou Jun and Ms. Cui Dandan; and guidance from the Department of Solid Wastes and Chemicals of the Ministry of Ecology and Environment, especially Mr. Qiu Qiwen (Director General), Ms. Ding Qiong and Mr. Zhou Zhiqiang (Deputy Director Generals), Mr. Wen Xuefeng, Mr. Sun Shaofeng and Mr. Zhang Long.

The guidance was compiled under supervision by FECO (Ms.

Zhou Guomei and Mr. Yu Lifeng), with technical guidance from Mr. Ren Yong, Mr. Chen Yang, Mr. Zheng Yang, Ms. Guo Chunxia and Mr. Jiang Wenbo.

The editorial board of the Chinese Edition is listed in the following page. This English Edition was developed based on the Chinese Edition edited by Mr. Ren Zhiyuan, Ms. Li Qiushuang, Mr. Zheng Zhe, Ms. Duan Lizhe and Ms. Qin Mingyu (executive editors and reviewers) and finally reviewed by Dr. Alessandro Flammini (UNIDO Headquarters).

Deep appreciation is extended to all the above.

Editorial Board

Training Guidance for the Management of Medical Waste during the COVID-19 Outbreak in China (Original Edition, in Chinese)

Consultants

Guo Jing Qiu Qiwen Zhou Guomei

Steering Group

Zhang Jieqing Zhou Zhiqiang Yu Lifeng

Experts

Chen Yang Duan Lizhe Hu Guoliang Jiang Wenbo

Guo Chunxia Li Chuanhua Sun Shaofeng Wang Yong

Wu Jinsong Xu Weibiao Yao Xiaoyong Zheng Yang

Chief Editors

Ren Yong Tang Jie Xu Yang Ren Zhiyuan Zheng Yang

Associate Editors

Li Qiushuang Jiang Wenbo Zheng Zhe Qin Mingyu

Chief Reviewers

Wu Jinsong Zheng Yang

Reviewers

Li Chuanhua Hu Guoliang Li Qiushuang

Writing Group

Guo Xiaojue Jiang Wenbo Li Qiushuang Li Yudong
Qin Mingyu Ren Zhiyuan Zhang Bo Zhang Hongxia
Zheng Zhe Zhong Jiangping

Preface

Medical waste refers to the waste with direct or indirect infectivity, toxicity and other hazards produced by medical and health institutions in medical treatment, prevention, health care and other related activities. Standardized management of medical waste is of great significance for providing a high quality, people-oriented medical service basis, safeguarding effectively people's health and environmental safety, accelerating the construction of ecological civilization, and promoting sustainable economic and social development.

In 2020, the pneumonia epidemic caused by novel coronavirus (COVID-19) infection is breaking out and the amount of medical waste in epidemic affected countries and regions is surging. The novel coronavirus has extremely high infectivity. It is crucial to ensure the proper disposal of medical waste with high infectivity and block effectively the secondary pollution caused by virus transmission, so as to safeguard people's health and safety and win the battle against the pandemic. It has become an extremely severe challenge on how to ensure the timely, safe, orderly and efficient disposal of medical waste during the epidemic.

In order to further strengthen the standardized management and disposal of medical waste during the epidemic, the Ministry of

Ecology and Environment (represented by the Foreign Environmental Cooperation Center, FECO) and the United Nations Industrial Development Organization (UNIDO) have swiftly cooperated, developed and implemented the project Emergency Response Assistance to the Outbreak of Coronavirus Disease 2019 in China, with support and coordination of the Ministry of Commerce. Under the guidance of the Ministry of Ecology and Environment and with support of the project, FECO coordinated the National Medical Waste Management Training Base (established by Shanghai Solid Waste Management Center and Shanghai Solid Waste Disposal Co., Ltd.) to jointly compile and complete the *Training Guidance for the Management of Medical Waste during the COVID-19 Outbreak in China*, which has obtained experts'strong support from the Technical Center for Solid Waste and Chemicals Management of the Ministry of Ecology and Environment, Basel Convention Asia Pacific Regional Center (School of Environment, Tsinghua University), University of Chinese Academy of Sciences, Henan Ecological Environment Department and other units.

This Guidance, based on China's current medical waste management system, combines closely with actual work practices and integrates highly with a series of national measures on medical waste management and emergency response during the epidemic. It covers the overview, classified collection and packaging of medical waste, internal transport, temporary storage, handover and registration, disposal, EHS management, supervision and management within medical institutions,

management regulations, management requirements for medical institutions, management requirements for medical waste disposal units as well as management and disposal cases during the pandemic. It is also a collection of experience and wisdom of such typical cities of Shanghai and Wuhan in the management and disposal of medical waste in the battle against COVID-19.

There are no national boundaries for the virus, and the epidemic endangers all races and nationalities. Mankind is a community with a shared future. Only through solidarity, cooperation and joint response can the international community overcome the epidemic ultimately. Solidarity and cooperation are the most powerful weapons. To fight the global war against COVID-19 resolutely, China is ready to share its beneficial practices in epidemic prevention and control with other countries and provide assistance within its capacity to the countries where the epidemic is spreading. In order to promote the transformation of China's experience in effective disposal and management of epidemic medical waste into a powerful weapon for other countries in need to fight the epidemic, FECO translated this Guidance into English. In this global battle against pandemic, it is believed the publication of this book will play a positive role in facilitating mankind to overcome the epidemic.

The compilation of this book is an exploratory work. It is the result of joint hard work of all the editors and reviewers with generous guidance from relevant leaders and experts. Appreciation is extended to

all of them. Due to the lack of time, certain deficiencies in this book are inevitable. Your valuable comments and suggestions are welcomed.

The Writing Group
May 2020

Contents

Part 1 Basic Concepts of Medical Waste

Part 2 Whole-process Management of Medical Waste

5 Medical Waste Disposal/27

6 EHS Management in Medical Waste Treatment Process/38

Part 1

Basic Concepts of Medical Waste

1 Definition and Overview of Medical Waste

1.1 What is the definition of medical waste?

In the *Regulations on the Management of Medical Waste*, medical waste refers to waste with direct or indirect infectivity, toxicity, and other hazards generated by medical and health institutions in medical treatment, prevention, health care, and other relevant activities. Household garbage generated by infectious patients or suspected infectious patients treated by medical and health institutions shall be managed and disposed of as medical waste. In addition, the management of directly or indirectly infectious waste, toxic waste and other hazardous waste generated in technical services for family planning, medical scientific research, teaching, autopsy, and other related activities shall be carried out in accordance with this Regulation. The management of anesthetic, psychotropic, radioactive, toxic drugs and related waste discarded by medical and health institutions shall be carried out in accordance with relevant laws, administrative regulations, rules and standards of the state.

1.2 What are the hazards that may be caused by medical waste?

(1) Infectious waste, pathological waste and medical sharp instruments may cause infectious diseases and even epidemic infectious

diseases, such as hepatitis B, hepatitis C, and AIDS.

(2) Genotoxic waste may cause gene mutation leading to hereditary diseases.

(3) Pharmaceutical waste and chemical waste may result in drug poisoning and chemical damage.

1.3 What are the objects that medical waste does harm to?

Anyone contacting (exposed, directly and indirectly contacting) medical waste who may be exposed to potential threats, including the staff of and visitors to medical institutions that produce medical waste, people contacting medical waste during home health care, staff for the collection, transfer and disposal of medical waste, staff for the supervision of medical waste environment and staff for health-epidemic prevention, the exposed population caused by ineffective management of medical waste, as well as the public exposed to toxic and harmful substance generated by the heat treatment of medical waste.

1.4 How to classify medical waste?

According to the *Classified Catalogue of Medical Waste*, medical waste is divided into 5 categories: infectious waste, pathological waste, sharps waste, pharmaceutical waste, and chemical waste. Refer to the following table for the characteristics and common components or names of various kinds of medical waste.

Category	Characteristics	Common Components or Names of Medical Waste	Schematic Diagram
Infectious waste	Infectious waste refers to the kind of medical waste that carries pathogenic microorganisms and has the hazard of leading to the spread of infectious diseases	Supplies contaminated by patients' blood, body fluid and excrement, including cotton ball, cotton swab, drainage cotton strip, gauze and other dressings; disposable sanitation supplies, disposable medical supplies and disposable medical devices; discarded bedding and clothing; other articles contaminated by patients' blood, body fluids and excrement	
		Household garbage generated by isolated infectious patients or suspected infectious patients treated by medical institutions	
		Pathogen culture medium and specimens, and preservation solution of strains and virus seed	
		Various discarded medical specimens	
		Discarded blood and serum	
		Used disposable medical supplies and disposable medical devices are regarded as infectious waste	

Category	Characteristics	Common Components or Names of Medical Waste	Schematic Diagram
Pathological waste	Pathological waste refers to human body waste generated during diagnosis and treatment as well as animal carcasses from medical experiments	Discarded human tissue and organs generated during surgery and other diagnosis and treatment processes	
		Animal tissue and carcasses from medical experiments	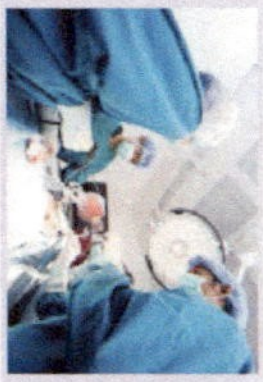
		Human tissue and pathological wax blocks discarded after pathological sections	
Sharps waste	Sharps waste refers to discarded medical sharp instruments that can stab or cut the human body	Medical needles and suture needles	
		All kinds of medical sharp instruments, including scalpels, surgical knives, skin preparation knives and surgical saws	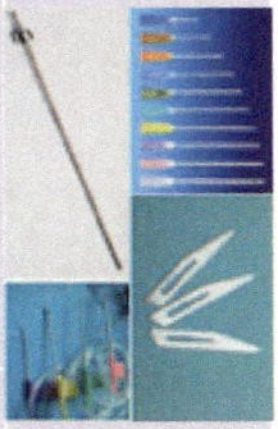
		Glass slides, glass test tubes, and glass ampoules	

Category	Characteristics	Common Components or Names of Medical Waste	Schematic Diagram
Pharma-ceutical waste	Pharmaceutical waste refers to discarded medicines that are expired, eliminated, deteriorated or contaminated	Discarded common medicines, such as antibiotics and OTC medicines	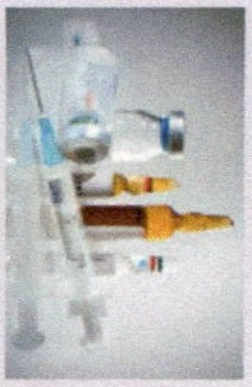
		Discarded cytotoxic drugs and genotoxic drugs, including carcinogenic drugs such as azathioprine, chlorambucil, chlornaphazine, cyclosporine, cyclophosphamide, amphetamine hydrochloride, semustine, tamoxifen, and thio-TEPA; suspected carcinogenic drugs, such as cis-platinum, mitomycin, adriamycin, and phenobarbital; immune inhibitor	
		Discarded vaccines and blood products	
Chemical waste	Chemical waste refers to discarded chemical articles that are toxic, corrosive, flammable and explosive	Discarded chemical reagents from medical imaging room and laboratory	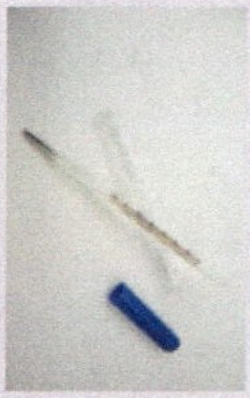
		Discarded chemical disinfectants such as peroxyacetic acid and glutaraldehyde	
		Discarded mercury sphygmomanometer and mercury thermometer	

1.5 What are the requirements of WHO for the management of medical waste?

In 1998, the World Health Organization (WHO) issued a report, which puts forward corresponding requirements for the management of medical waste in terms of national legislation and management action plans, internal management of medical institutions, medical waste reduction measures, collection, classification, packaging, transport, treatment and disposal of medical waste, preferentially-supported activities, and development strategies. The report also points out that safe and reliable medical waste disposal methods are the most important; the effective management and disposal of medical waste objectively requires the cooperation and coordination of the whole society; establishing a complete set of laws and regulations, carrying out personnel training and raising public awareness are vital for the successful implementation of the management and disposal of medical waste.

In 2014, WHO issued a revised edition of the *Safe Management of Wastes from Health-care Activities*, which provides guidance for the management of medical waste under normal and emergency situations. In December 2014, WHO released a series of guidance documents on Ebola virus, providing guidance on the simple treatment and disposal of medical waste. Among them, the disposal of medical waste in the emergency state of natural disasters or war conflicts is stipulated as follows: "When resources are unavailable, medical waste can be treated and disposed of by the following minimum methods: burying in the

pit or trench on site, burning in a simple two-chamber incinerator, packaging sharps waste or a small amount of drugs and then burying on site or burying in the special area of municipal landfills, burning in a high-temperature industrial incinerator (provided there is a safe vehicle of transport), and sterilizing infectious waste and sharps waste with a small high-pressure sterilizer." Both the *Infection Prevention and Control during Health care for probable or confirmed cases of Middle East Respiratory Syndrome Coronavirus (MERS-CoV) Infection -Interim Guidance issued* by WHO in June 2015 and the *Infection Prevention and Control During Health Care When Novel Coronavirus (COVID-19) Infection Is Suspected issued* by WHO on January 25, 2020 have pointed out that "Medical waste should be managed in accordance with safe routine procedures".

1.6 What are the requirements of the Stockholm Convention on Persistent Organic Pollutants for the management of medical waste?

Dioxins are the primary unintentionally-produced persistent organic pollutants required to be controlled by the Stockholm Convention on Persistent Organic Pollutants, and medical waste incineration is one of the important sources of dioxins emissions. The Convention requires parties to adopt the Best Available Techniques and the Best Environmental Practices (BAT/BEP) to reduce dioxins emissions. General BAT/BEP measures related to the management of medical waste in Annex C of the Convention include: (1) Reduction; (2) Stop open

burning or simple burning; (3) Give priority to incineration alternative measures to be adopted in newly-built facilities; (4) Reduce pollution emissions during disposal.

The BAT/BEP Guideline (Guidelines on best available techniques and provisional guidance on best environmental practices relevant to Article 5 and Annex C of the Stockholm Convention on Persistent Organic Pollutants) highlights that although incineration technology is the most mature technology for medical waste disposal, it also has relatively great environmental risks, and non-incineration alternative technologies such as high-temperature steam, microwave and chemical treatment are encouraged to be adopted. Specific requirements for the management and disposal of medical waste are as follows: (1) Improve government legislation; (2) Establish a classified catalogue of medical waste; (3) Promote source reduction, strictly distinguish household garbage from medical waste, and reduce the use of disposable medical supplies; (4) Reduce the use of chlorine-containing medical supplies; (5) Establish an effective medical waste management system; (6) Adapt classification to disposal methods; (7) Build new medical waste incinerator only when necessary to reduce the release from incineration; (8) Advocate the recycling of resources as much as possible on the premise of meeting public health security; (9) Encourage the use of advanced technologies and eliminate obsolete disposal technologies; (10) Support the development and promotion of waste reduction technology research, and promote the development and application of

environmentally friendly disposal technologies.

1.7 What are the requirements of the Basel Convention for the management of medical waste?

The Basel Convention on the Control of Transboundary Movements of Hazardous Wastes and Their Disposal points out that medical waste is infectious and one of the important components of hazardous waste. The Convention lists medical waste in Annex I, aiming to realize environmentally sound management, protect human health, and reduce its damage to the environment. In order to promote the implementation of the Convention, the Secretariat of the Basel Convention has prepared the *Technical Guidelines for the Environmentally Sound Management of Biomedical and Health Care Waste* (Y1; Y3) in accordance with the Convention.

The Guidelines put forward corresponding requirements on how to avoid and prevent waste generation, how to classify, collect, mark and transfer the waste, how to transport and temporarily store the waste, and how to recycle the waste. In terms of specific disposal methods, the Guidelines also put forward corresponding suggestions and recommended disposal methods for different waste types, such as high-temperature steam treatment, chemical disinfection treatment, microwave irradiation treatment, and incineration treatment.

1.8 What laws and regulations promulgated in China are related to the management of medical waste ?

The management of medical waste in China is mainly based on

the *Regulations on the Management of Medical Waste* promulgated in 2003. The Regulation is applicable to the collection, transport, storage, disposal, supervision and management of medical waste. The Regulation stipulates that the health department and the environmental protection department are the two main supervision departments - the former is responsible for supervising the prevention and treatment of disease in the whole process of management, and the latter is in charge of supervising the prevention and control of environmental pollution in the whole process of management. In order to perform their functions and powers, the two departments have respectively formulated the *Measures for the Administration of Medical Waste in Medical and Health Institutions* and *the Technical Specifications for Concentrated Disposal of Medical Waste*, and to urge the management objects to assume legal responsibilities, the *Administrative Punishment Measures for the Management of Medical Waste* is formulated. China has also issued the *Classified Catalogue of Medical Waste*, various disposal technical specifications and local legal regulations.

the Regulations on the Management of Medical Waste, which was promulgated in 2003. The regulations [illegible] collection, transport, storage, disposal, supervision and management of medical waste. The Regulations stipulate that the health administrative department and the environmental protection department are the two main supervision departments. The former is responsible for supervising the prevention and control of disease in the whole process of medical waste management, and the latter is responsible for supervising the prevention and control of environmental pollution in the whole process of medical waste management. In order to perform their duties and powers, the two departments have jointly formulated the Measures for the Administrative [illegible] of Medical Waste in Medical and Health Institutions and [illegible], Catalogue of Medical Waste, and [illegible] the main [illegible] to [illegible] [illegible] Medical Waste [illegible] the [illegible] of Medical Waste [illegible] technical specifications [illegible].

Part 2

Whole-process Management of Medical Waste

2 Classified Collection and Packaging of Medical Waste

2.1 Can different types of medical waste be mixed and collected together?

Medical and health institutions shall implement classified management of medical waste according to the *Classified Catalogue of Medical Waste*. Schematic diagrams or written descriptions of classified collection methods of medical waste shall be set up at the place where medical waste is generated to make it clear that infectious waste, pathological waste, sharps waste, pharmaceutical waste and chemical waste cannot be mixed together and collected. A small amount of pharmaceutical waste can be mixed into infectious waste, but it shall be indicated on the label. It is prohibited to mix medical waste with other wastes and household garbage.

Special reminder: there are special requirements for high-risk waste, medical waste and excrement generated by patients with infectious diseases. The purpose of classified collection of some types of medical waste is to send them to specialized institutions for disposal, instead of to centralized disposal units for mixed disposal. The management of anesthetic, psychotropic, radioactive, and toxic drugs and related waste discarded by medical and health institutions shall be carried out in accordance with relevant laws, administrative regulations, rules and standards of the state.

2.2 What belongs in high-risk medical waste? How to deal with it before collection?

The *Regulations on the Management of Medical Waste* stipulates that high-risk waste such as pathogen culture medium and specimens, preservation solution of strains and virus seed in medical waste shall be treated by pressure steam sterilization or chemical disinfection treatment at the place of generation, and then collected and treated as infectious waste.

2.3 How shall the medical waste generated by infectious patients be collected?

Household garbage generated by infectious patients or suspected infectious patients treated by medical and health institutions shall be managed and disposed of as medical waste. Double-layer packaging material shall be used and it shall be sealed in time.

2.4 How shall chemical medical waste be collected and disposed of?

Chemical medical waste shall be placed in special packaging container, and it is forbidden to mix it into other types of medical waste. Flammable and explosive chemical waste shall be collected and transported separately. Medical and health institutions shall, in accordance with relevant procedures of hazardous waste disposal, hand it over to hazardous waste disposal units with corresponding disposal capabilities for disposal. Centralized disposal units for medical waste

shall assist in confirming relevant disposal destination.

2.5 What are the requirements for the frequency and time limit for medical waste collection?

Medical waste generation units shall collect medical waste generated in their own units no less than once a day; medical waste generated in medical activities such as medical tour and on-site first aid shall be collected immediately after the end of medical activities.

2.6 What requirements shall be met by classified collection points for medical waste?

Classified collection points refer to temporary storage points receiving medical waste sent by medical personnel in various medical positions. Classified collection points are set up in outpatient (emergency) clinic, medical departments, medical technology departments and wards (inpatient area) for the classified collection, packaging and management of medical waste before delivery to the temporary storage warehouse of the institution. Classified collection points for medical waste in small medical institutions such as clinics are usually set up together with temporary storage warehouse and temporary storage cabinets.

2.7 What emergency measures shall be taken when a medical waste accident happens in medical institutions?

Identify the type and amount of medical waste lost, leaked and

diffused, the time, affected scope and severity of the accident. Organize relative personnel to clean the leaked and diffused site according to the emergency plan.

When cleaning the area polluted by medical waste, the effect on patients, medical staff, other personnel on site and the environment shall be minimized.

Take proper safety disposal measures, conduct disinfection or other harmless treatment measures to the leakage and polluted areas and supplies, and lock down the polluted area if necessary.

When disinfecting the area polluted by infectious waste, the disinfection work shall be started from the less polluted area to the severer area, and the tools that may be polluted shall also be disinfected.

The staff shall carry out their work after ensuring their health and protection. After the disposal is done, the medical institution shall investigate the cause of the event and take effective preventive measures to prevent similar events.

The emergency report of the medical institution mainly contains the time, place and brief overview of the accident; the type and amount of the medical waste lost, leaked and diffused; possible cause of the accident; the hazard and impact caused by the accident; and emergency disposal measures and results.

2.8 What are the key requirements for packaging containers specific to medical waste?

The *Standard of Packaging Bags, Containers and Warning*

Symbols Specific to Medical Waste (HJ 421-2008) stipulates the technical requirements for packaging bags, sharps boxes and recycle boxes (bins) specific to medical waste, corresponding test methods and inspection rules, and warning symbols for medical waste.

When the volume of medical waste reaches 3/4 of the full size of the packaging material or container, an effective sealing method shall be used to make the sealing of the packaging material or container tight.

Medical waste placed in the packaging material or container shall not be unpacked and taken out again during the whole process of collection, transport, storage and disposal.

Sharps medical waste including medical needles, suture needles, all kinds of medical sharp instruments, glass slides, glass test tubes, and glass ampoules, shall be contained in disposable special rigid containers, i.e. sharp boxes.

For medical waste treated by high-temperature thermal treatment technology, PVC materials shall not be used for packaging bags and sharp boxes.

When the outer surface of the packaging material or container is contaminated by infectious waste, the contaminated part shall be disinfected or an extra layer of packaging shall be used.

3 Transport and Temporary Storage of Medical Waste within Medical Institutions

3.1 What are the requirements for transport time and route of medical waste?

The time and route for transporting medical waste within the hospital shall be relatively fixed. The selection of transport route shall be based on the principle of minimum flow of people and goods or relatively remote location. Transport time shall avoid peak hours of diagnosis and treatment. In the process of transport, the transporter shall not leave the transfer vehicle. Medical waste shall be transported to designated temporary storage sites. Transport time and route cannot be changed at random.

3.2 Must the transport tools be specific to medical waste?

Transport tools (including transport vehicles and containers) specific to medical waste transport shall be anti-leakage, spill-proof, free of sharp edges and corners, and easy for loading and unloading and cleaning, and warning symbols and written instructions for medical waste shall be printed (sprayed) on the outer surface. After the daily transport is over, the delivery tools shall be cleaned and disinfected in a timely manner. Medical waste shall not be transported by special tools

that have not been disinfected or cleaned.

3.3 What shall be checked before transporting medical waste?

Before transporting medical waste, transporters shall check whether special packaging bags and sharps boxes are used for the classification and collection of medical waste, whether the label content on each medical waste packaging bag and sharps box meets the requirements, whether the packaging bags are damaged and whether the sealing is tight. If it is found that it does not meet the requirements, correction requests shall be put forward to relevant personnel responsible for the management of medical waste, and medical waste can be transported only after the requirements are met through correction. Medical waste that fails to meet the requirements shall not be transported to temporary storage sites.

3.4 What shall be paid attention to when transporting medical waste?

It is forbidden to discard medical waste during transport. Article 18 of the *Measures for the Administration of Medical Waste in Medical and Health Institutions* stipulates that when transporting medical waste, it is necessary to prevent damage to packaging materials or containers, as well as loss, leakage and diffusion of medical waste, and prevent medical waste from directly contacting the body.

3.5 What are the requirements for the temporary storage facilities and equipment for medical waste?

Medical and health institutions shall not store medical waste in the open air. They shall establish temporary storage facilities and equipment for medical waste which shall meet the following requirements: (1) They shall be far away from the medical area, food processing area, personnel activity area and household garbage storage place, and also should facilitate the access of medical waste transport personnel, transport tools and vehicles; (2) They shall be tightly shut, and special (part-time) staff shall be assigned for management and non-staff shall be prevented from contacting medical waste; (3) There shall be safety measures against rats, mosquitoes, flies and cockroaches; (4) There shall be measures against leakage and rain erosion; (5) They shall be convenient to clean and disinfect; (6) Direct sunlight shall be avoided; (7) There shall be obvious medical waste warning symbols and signs writing "禁止吸烟、饮食" ("No Smoking or Eating" in English).

3.6 How shall medical waste be stored if it is difficult to set up an independent temporary storage warehouse for medical waste?

Medical and health institutions that do not have inpatient beds, such as outpatient departments, clinics, medical education institutions and scientific research institutions, shall set up temporary storage cabinets (boxes) specific to medical waste when it is difficult for them to

set up an independent temporary storage warehouse for medical waste, and the following requirements shall be met: the temporary storage cabinets (boxes) specific to medical waste must be separated from the place where the household garbage is stored with measures against rain and scattering, and they must meet fire safety requirements; after the classified and packaged medical waste is placed in the turnover box, it shall be placed in the special temporary storage cabinet (box). The cabinet (box) shall be sealed, and safety measures shall be taken, such as locking and fixing, so that irrelevant personnel cannot move it, and warning symbols shall be set outside; refrigerated cabinets (boxes) can be used as special temporary storage cabinets (boxes) for medical waste; the cabinets (boxes) can also be made of metal or rigid plastic with certain strength and anti-leakage performance.

3.7 How long is the temporary storage period of medical waste in medical institutions?

Medical waste shall be cleared every day as far as possible. If this cannot be achieved and the local maximum temperature is higher than 25℃ , medical waste shall be temporarily stored at a low temperature (lower than 20℃) for a maximum of 48 hours.

Temporary storage of pathological waste, except for that it shall meet the requirements for temporary storage of medical waste, shall also ensure low-temperature or antiseptic condition.

3.8 What are the requirements for the cleaning and disinfection of temporary storage warehouse for medical waste?

After medical waste is transferred out, temporary storage sites and facilities shall be cleaned and disinfected in time.

The temporary storage warehouse for medical waste shall be disinfected and flushed every day after the waste is removed and transported. Flushing liquid shall be discharged into the disinfection and treatment system for medical waste water in medical and health institutions. Temporary storage cabinets (boxes) for medical waste shall be disinfected once a day.

4 Handover and Transfer of Medical Waste

4.1 How shall medical waste in medical institution be transferred?

Within the medical institution, medical waste shall be successively handed over from the medical personnel in the position, to the classified collection and management personnel of each department, to the special medical waste transfer personnel of the institution, and to the temporary storage and management personnel of the institution, and handover records shall be made for each link. Finally it shall be handed over to the collection personnel of the centralized disposal unit.

4.2 What is the content of medical waste registration (standing book)? How long shall the registration document be kept?

Medical and health institutions and centralized medical waste disposal units shall register information of medical waste respectively, and the content of registration shall include items such as the source, category, weight or quantity, handover time, disposal methods, final destination, and signature of the handlers. Registration document shall be kept for at least 3 years for local environmental protection department and health department to check.

4.3 What shall be registered for the handover of medical waste from medical institutions to centralized disposal units?

For medical waste handed over by medical and health institutions for disposal, the Duplicate Tables for the Transfer of Hazardous Waste (specific to medical waste) shall be filled in. After inspection and verification, if the packaging and identification of the received medical waste comply with the regulations and the type and weight are consistent with the items contained in the duplicate tables for the transfer, the handover personnel of the medical and health institution and the transport personnel of the centralized disposal unit shall sign and confirm the duplicate tables respectively. One Registration Card for Medical Waste Transport is for one vehicle, which shall be filled in and signed

by the personnel responsible for the management of medical waste of the medical and health institution during the handover. When medical waste is transported to the disposal unit, the receiving personnel of the centralized disposal unit shall sign after confirming that the quantity of medical waste filled in the Registration Card is true and accurate.

4.4 What requirements shall medical waste transport vehicles meet?

Medical waste transport vehicles shall comply with the provisions of the state on the transport and management of dangerous goods, have obvious medical waste identification, and meet the requirements of anti-seepage, spill-proof and other environmental protection and health requirements. After the vehicle is used, it shall be disinfected and cleaned in a timely manner in the centralized disposal site for medical waste. Special vehicles transporting medical waste shall not transport other supplies.

4.5 What shall medical waste transport vehicles be equipped with?

Transport vehicles shall be equipped with: a copy of the *Technical Specifications for Concentrated Disposal of Medical Waste (Trial)*, Duplicate Tables for the Transfer of Hazardous Waste (specific to medical waste), Registration Card for Medical Waste Transport, transport route map, communication equipment, a name list of medical waste generating units with contacts of management personnel, accident

emergency plan and a name list of contact units with phone numbers of relevant personnel, tools for collecting medical waste, disinfection equipment and medicines, spare bags and sharps boxes specific to medical waste, and spare personnel protective supplies.

4.6 What are the rules for the cleaning and disinfection of medical waste transport vehicles?

After each time of transport of the special medical waste transport vehicle, the inner wall of the van shall be disinfected in the disposal unit and the van shall be sealed for at least 30 minutes after spraying disinfectant. Medical waste transport vehicles shall be cleaned at least once every 2 days, or immediately after the inner wall or/and the outer surface of the van is contaminated. It is forbidden to clean medical waste transport vehicles in cleaning places for general vehicles.

4.7 Can medical waste be transported by water/railway/air?

It is forbidden to mail medical waste. It is forbidden to transport medical waste by railway or air. If there is a land route, it is forbidden to transport medical waste by water; if there is no land route and medical waste must be transported by water, it shall be approved by the competent department of environmental protection administration of the people's government at or above the municipal level, and strict environmental protection measures shall be taken before it can be transported by water. It is forbidden to transport medical waste on the

water body in protection area for drinking water source.

5 Medical Waste Disposal

5.1 What qualifications shall a centralized disposal unit for medical waste obtain?

Any unit engaged in centralized disposal of medical waste must obtain the hazardous waste business license. Unit without license shall not engage in relevant activities.

5.2 What are the disposal methods of medical waste?

At present, the mainstream treatment and disposal methods of medical waste include high-temperature incineration, high-temperature steam disinfection, microwave disinfection, chemical disinfection, and high-temperature dry heat disinfection.

5.3 What are the principle and characteristics of incineration disposal technology for medical waste ?

Medical waste incineration disposal process uses high-temperature heat treatment to make the organic components in the medical waste undergo oxidation/decomposition reactions to reduce waste and make medical waste harmless, mainly including pyrolysis incineration technology and rotary kiln incineration technology. It is suitable for the disposal of infectious, injury, pathological, chemical and pharmaceutical medical waste.

Pyrolysis incineration disposal facility

Rotary kiln incineration disposal facility

Rotary kiln incineration system consists of a rotary kiln and a secondary combustion chamber. Medical waste is burned in the rotary kiln with oxygen-rich combustion to complete the process of drying, incineration, burnout and cooling. The boiling steam and combustion gas enter the secondary combustion chamber for complete combustion.

Pyrolysis incineration system consists of a pyrolysis gasifier and a

secondary combustion chamber. The pyrolysis gasifier uses the principle of hypoxia pyrolysis to supply an insufficient amount of combustion air, so that medical waste is cracked into short-chain flammable gas in a certain temperature range and under hypoxic conditions. The cracked gas enters the secondary combustion chamber for complete combustion.

5.4 What are the main points for the operation and management of incineration facility ?

It shall ensure stable combustion in the incinerator during feeding. The feeding inlet shall be kept airtight. It shall ensure smooth feeding and prevent waste from blocking. Medical waste packaging bag shall be kept intact before entering the furnace. The feeding system shall be in negative pressure state to prevent harmful gases from overflowing.

During the normal operation of the medical waste incineration disposal facility, the incinerator shall be in a negative micro-pressure combustion state; control the flue gas temperature in the secondary combustion chamber to be no lower than 850°C (the incineration of chemical waste and pharmaceutical waste requires the flue gas temperature in the secondary combustion chamber to be no lower than 1,100°C), and the flue gas residence time is no less than 2 s. The combustion efficiency of the equipment shall be no lower than 99.9%, and the thermal reduction rate of the incineration residue shall be less than 5%. The oxygen content in the flue gas at the outlet of the incinerator shall be controlled within the range of 6% ~ 10% (dry gas).

Medical waste shall be completely incinerated, and the temperature, residence time and turbulent conditions of the flue gas in the combustion chamber shall be strictly controlled. Rapid cooling measures shall be taken for the high-temperature flue gas generated by the combustion of medical waste, and the residence time of the flue gas at 200 ～ 500°C shall be less than 1s.

The slag generated by incineration can be sent to a designated municipal waste sanitary landfill for landfill disposal. Residues such as incineration fly ash and activated carbon adsorbing dioxins and other harmful components shall be managed as hazardous waste.

5.5 What are the principle and characteristics of high-temperature steam treatment of medical waste ?

The high-temperature steam treatment process of medical waste uses the latent heat released by water vapor to cause protein denaturation and coagulation of pathogenic microorganisms to disinfect the medical waste, mainly including two types of process: steam treatment followed by crushing and steam treatment and crushing at the same time. The process has the characteristics of low investment, low operating cost, simple operation and less environmental pollution, and is suitable for the treatment of infectious and injury medical waste.

High-temperature steam treatment facility

5.6 What are the main points for the operation and management of high-temperature steam treatment facility?

Feeding link: try to avoid the feeding container (or feeding vehicle) from touching human body directly if manual feeding is adopted. The medical waste filling in the feeding container (or feeding vehicle) shall be in appropriate loose status and the largest loading capacity shall not surpass 70% of the volume of the disinfection room.

High-temperature steam treatment link: the indoor temperature of the disinfection room shall be no lower than 134°C, its pressure shall be no lower than 220kPa (gauge pressure) and the treatment time shall be no less than 45 min. Ensure the steam supply capacity, temperature and pressure. Maintain the steam supply pressure within 0.3 ～ 0.6 MPa. Waving capacity shall be no more than 10% and non-condensable gas

(volume fraction) no more than 5%. Overheating shall not exceed 2°C. After high-temperature steaming and drying, the water content of medical waste residues shall not exceed 20% of the total weight.

Crushing link: Maintenance of crushing equipment for medical waste shall be carried out regularly. The particle size of crushed medical waste shall be maintained as less than 5cm. When crushing medical waste that has not been sterilized yet, leakproofness of the operation workshop shall be ensured and the crushing shall be conducted in negative pressure state.

Detection of disinfection effect: use bacillus stearothermophilus as indicating bacterium to make sure that the killing log value is no less than 4 to meet the requirement for disinfection effect.

5.7 What are the principle and characteristics of chemical disinfection treatment of medical waste ?

Chemical disinfection of medical waste is to well mix the crushed medical waste and chemical disinfectant (such as lime powder, sodium hypochlorite, calcium hypochlorite, chlorine dioxide, etc.) and let it stay long enough so the organic substance can be resolved and infectious bacteria can be killed or inactivated during the disinfection.

The process has the advantages of low investment at a time, low operating cost, being economically applicable and less pollution to the environment. It is suitable for the treatment of infectious and injury medical waste.

Chemical disinfection treatment facility for medical waste

5.8 What are the main points for the operation and management of chemical disinfection treatment facility ?

Before operating the chemical disinfection treatment unit, the crushing unit shall be ensured to operate in airtight and negative pressure state and the air in it shall be ensured to be emitted after disinfection. Workers shall get enough disinfection before making inspections.

Chemical disinfectant's killing efficiency of microorganisms relates to their contact time, temperature, chemical disinfectant's concentration, pH value (acid and alkaline environment) and the quantity and type of the microorganisms. In order to obtain the effect of inactivation, chemical disinfectant must have adequate contact reaction time with the disinfection surface of medical waste and enough input volume of disinfectant must be ensured.

Currently, the frequently used method in China is dry chemical disinfection. Lime powder is the main ingredient for sterilization with

the purity being 88%~95% and the contact reaction time shall be longer than 120 min. The dosage of input volume of dosage (lime powder/ medical waste) shall be over 0.075 kg/kg while the pH of the strong alkaline environment under reaction control shall be 11.0 ~ 12.5.

Effect detection: Bacillus subtilis spores are adopted as indicating bacterium to ensure that the killing log value of microorganisms is no less than 4 to meet the requirement for disinfection effect.

5.9 What are the principle and characteristics of microwave disinfection treatment of medical waste ?

Microwave disinfection treatment of medical waste adopts the heat generated from microwave vibration of water molecules to inactivate infectious bacteria and disinfect medical waste. The technology has the characteristics of broad bactericidal spectrum, no residue, good effect in deodorization, and cleanliness. It is applicable for the treatment of infectious and injury medical waste.

Microwave disinfection treatment facility for medical waste

5.10 What are the main points for the operation and management of microwave disinfection treatment facility?

Feeding link: Once the equipment starts working, feeding shall keep unobstructed and prevent blocking. Package volume of medical waste shall match the size of feeding inlet to ensure medical waste and its package enter the process unit under intact condition. When the feeding equipment starts to operate, air-extractor shall be turned on at the same time to maintain the system operating under negative pressure.

Microwave disinfection link: The frequency of microwave shall be (915±25) MHz or (2 450±50) MHz. The temperature shall be no lower than 95°C and the working time shall be no less than 45 min. If pressure is applied during the treatment, the temperature of the materials treated by microwave shall be lower than 170°C to avoid secondary pollution caused by decomposition of chlorine-containing compounds such as plastics in medical waste. Under the combined action of steam and microwave, when the temperature is no lower than 135°C, the action time is no less than 5 min. Check the safety protection procedure for microwave disinfection equipment before operation to avoid electromagnetic wave leakage.

Crushing link: Maintenance of crushing equipment for medical waste shall be carried out regularly. The particle size of crushed medical waste shall be maintained as less than 5cm. The crushing process shall be conducted under airtight and negative pressure state.

Effect detection: Bacillus subtilis spores are adopted as indicating bacterium to ensure that the killing log value of microorganisms is no less than 4 to meet the requirement of disinfection effect.

5.11 What are the principle and characteristics of high-temperature dry heat treatment of medical waste ?

The principle of high-temperature dry heat treatment of medical waste is to expose medical waste in negative-pressure and high-temperature environment after high intensity grinding and let it stay for some time, and apply precise transferring program to transfer heat effectively to medical waste to be treated in order that pathogenic microorganisms undergo protein denaturation and coagulation which results in the death of pathogenic microorganisms of medical waste and makes medical waste harmless, thus realizing safe disposal. High-temperature dry heat treatment technology is suitable for the treatment of infectious and injury waste.

High-temperature dry heat treatment facility for medical waste

5.12 What are the main points for the operation and management of high-temperature dry heat treatment facility?

Feeding link: Once the equipment starts working, feeding shall keep unobstructed and prevent blocking. Package volume of medical waste shall match the size of feeding inlet to ensure medical waste and its package enter the process unit under intact condition. When the feeding equipment starts to operate, air-extractor shall be turned on at the same time to maintain the system operating under negative pressure.

Dry heat treatment link: The pressure in the sterilizer shall be 300Pa, which is close to the vacuum; the temperature in the sterilizer shall be 180 ~ 200°C. The process time shall be no less than 20 min and the stirring device shall operate in a speed of no lower than 30r/min. Man-made interference shall be avoided during the operation of the dry head treatment equipment to prevent the equipment from man-made shutdown before completing the disinfection of medical waste.

Crushing link: Maintenance of crushing equipment for medical waste shall be carried out regularly. The particle size of crushed medical waste shall be maintained as less than 5cm. The crushing process shall be conducted under airtight and negative pressure state.

6 EHS Management in Medical Waste Treatment Process

6.1 What aspects does EHS management include?

EHS is the abbreviation of Environment, Health and Safety. EHS management system is the integration of environmental management system (EMS) and occupational health and safety management system (OHSMS).

6.2 What pollution control standards shall medical waste treatment process meet?

In medical waste treatment process, the impact on the environment caused by environmental hazard factors such as water, air, sound and slag shall be strictly controlled. The emission standards of various pollutants must conform to the national, local and industry-related control standards for waste water, waste gas, noise and solid pollutants, such as *Integrated Emission Standard of Air Pollutants, Standards for Pollution Control on Hazardous Waste Incineration, Emission Standards for Odor Pollutants, Integrated Wastewater Discharge Standard, Wastewater Quality Standards for Discharge to Municipal Sewers and Emission Standard for Industrial Enterprises Noise at Boundary.*

6.3 Does the centralized disposal unit for medical waste need to install an automatic monitoring system for pollutant emission?

Centralized disposal units for high-temperature incineration of

medical waste shall install a continuous automatic monitoring system for flue gas emission from fixed pollution sources. The monitoring items shall at least include sulfur dioxide, nitrogen oxides, particulate matter, carbon monoxide, oxygen content, hydrogen chloride, flue gas temperature, flue gas pressure, flue gas flow velocity or flow rate, and moisture content of flue gas. Automatic monitoring data can be used for environmental law enforcement.

6.4 Is medical waste incineration residue a kind of hazardous waste or general solid waste?

Incineration residues mainly include fly ash generated by dust removal equipment and slag generated from incineration. Fly ash is a kind of hazardous waste and must be collected and stored in a sealed manner, and solidified for landfill disposal in accordance with GB18598 *Standard for Pollution Control on the Hazardous Waste Landfill*; slag can be sent to household garbage landfills.

6.5 Where and what kind of warning symbols shall disposal units for medical waste set up?

Disposal units for medical waste shall set up hazardous waste warning symbols and medical waste warning symbols at the entrance and exit of the factory area, temporary storage facilities and disposal sites.

6.6 How shall medical waste disposal practitioners carry out occupational health protection?

Medical and health institutions and centralized medical waste

disposal units shall take effective occupational health protective measures, and provide necessary protective equipment for personnel engaged in medical waste collection, transport, storage, and disposal as well as management. Operators shall wear protective gloves, masks, work clothes, boots and other protective supplies during the operation. In case of danger caused by liquid or molten material spillage, operators shall also wear goggles; regular health checks shall be conduct; when necessary, relevant personnel shall be immunized to prevent them from health damage.

6.7 How shall medical waste collection and transport personnel carry out occupational health protection?

Transport personnel shall wear protective gloves, masks, work clothes, boots and other protective supplies during the transport; transport personnel shall conduct physical examination: 2 times per year, and take preventive immunization if necessary.

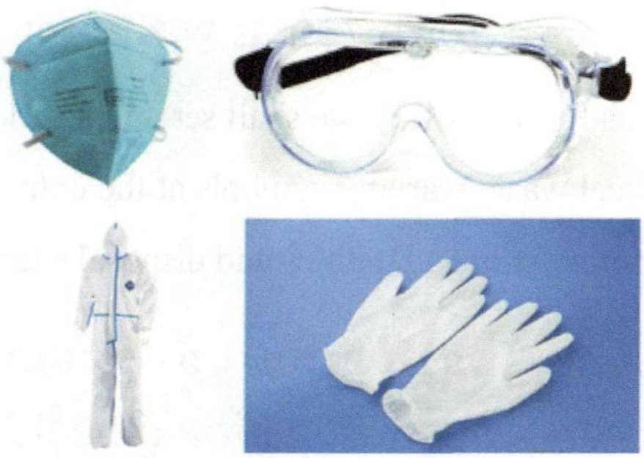

Common Personal Protective Equipment

6.8 How shall medical waste disposal practitioners disinfect themselves and the equipment and protective supplies they use?

Disinfection of hand and skin: soak or wipe hands with 0.5% iodophor solution (containing 5,000 mg/L of effective iodine) for 1 ~ 3 minutes. Disinfection of instruments and equipment: wipe and disinfect with 0.2% peracetic acid, 500 mg/L chlorine dioxide, or disinfectant solution of available chlorine or bromine. Disinfection of protective supplies: heat-resistant supplies can be disinfected with circulating steam for 20 ~ 30 minutes or with pressure steam at 121°C for 20 ~ 30 minutes, and moisture-resistant supplies (including protective glasses) can be soaked in disinfectant with available chlorine content of 1,000 mg/L for 30 minutes.

6.9 What disinfection method shall be used for medical waste transport tools?

Disinfection of the surface and environment of transport vehicles, turnover boxes (bins), storage facilities and other related supplies: sprinkle, spray or wipe with 0.2% peracetic acid solution or disinfectant with available chlorine content of 1,000 ~ 2,000 mg/L, and wash off after 60 minutes. Floor disinfection: use chlorine dioxide with content of 500 ~ 1,000 mg/L, or disinfectant with available chlorine or available bromine content of 1,000 ~ 2,000 mg/L to spray or mop the floor.

6.10 What are the roles involved in the whole process of supervision and management of medical waste?

Supervision and management of medical waste involves local people's government, hygiene and health management department, ecological environment department, medical and health institutions, centralized disposal units for medical waste, and transport department.

6.11 How to carry out supervision and inspection of medical waste in medical institutions?

Focus on the supervision and inspection of the establishment of responsibility system for medical waste management at all levels and types of medical institutions, the staffing of monitoring departments and specialized (part-time) management personnel, the implementation of management system and emergency plan, the development of occupational health and safety protection and training for staff, the classified collection, internal transport, handover and registration, temporary storage, and centralized disposal of medical waste to prevent the loss, leakage and spread of medical waste.

6.12 How to carry out supervision and inspection of centralized medical waste disposal units?

According to the regulation and requirements for the supervision and management of operation of centralized incineration facilities for

medical waste, the setting of hazardous waste identification marks, the formulation of hazardous waste management plans, the implementation of management systems such as hazardous waste declaration and registration, duplicate tables for transfer, business license, emergency plan filing, as well as whether the storage, utilization, and disposal of hazardous waste conform to relevant standards and specifications shall be checked item by item.

medical waste, the [illegible] of hazardous waste identification tables, the formulation of hazardous waste management plans, the implementation of management systems such as hazardous waste declaration and registration, [illegible] tables for transfer, business license, emergency [illegible] filing, as well as whether the storage, utilization, and disposal of hazardous waste conform to relevant standards and specifications should be checked item by item.

Part 3

Management of Medical Waste during an Epidemic

7 Regulations on the Management of Medical Waste during an Epidemic

7.1 What normative documents did China issue on the management of medical waste during the COVID-19 epidemic?

During the novel coronavirus pneumonia epidemic (referred to as the "COVID-19 epidemic"), the General Office of National Health Commission of the people's Republic of China issued the *Notification on the Management of Medical Waste in Medical Institutions during the Novel Coronavirus Pneumonia Epidemic*; the Ministry of Ecology and Environment of the people's Republic of China issued the *Notification on the Environmental Management of Medical Waste during the Novel Coronavirus Pneumonia Epidemic* and the *Technical Guidelines on the Emergency Disposal of Medical Waste from the Treatment of COVID-19 (Trial)*; and some local governments also issued policies for the management of medical waste. The National Health Commission requires to strengthen the management of institutions and personnel that provide logistics services, organize training, and urge them to master the basic requirements for the management of medical waste, genuinely perform their duties and dispose of the medical waste timely.

7.2 What is COVID-19 epidemic-related medical waste?

According to the *Notification on the Management of Medical*

Waste in Medical Institutions during the Novel Coronavirus Pneumonia Epidemic (GWBYYH [2020] No.81) issued by the National Health Commission of the people's Republic of China, medical waste of COVID-19 epidemic has been clarified: waste generated in fever clinics and impatient areas (wards) of medical institutions during the diagnosis and treatment of COVID-19 patients and suspected COVID-19 patients, including medical waste and household garbage, shall be classified and collected as medical waste.

7.3 How shall the technical route for emergency disposal of medical waste be determined during the COVID-19 epidemic?

First, centralized medical waste disposal units shall give priority to the disposal of epidemic-related medical waste. Second, professional mobile medical waste disposal facilities and non-professional facilities for medical waste disposal can be selected as alternative resources for the emergency disposal of medical waste. Professional mobile medical waste disposal facilities include incineration, high-temperature steam disinfection, microwave disinfection, and chemical disinfection; non-professional disposal facilities for medical waste refer to solid waste disposal facilities adopting high-temperature incineration technology, such as hazardous waste incineration facilities (with incineration temperature higher than 1,100°C), household garbage incineration facilities (with incineration temperature higher than 850°C), and

industrial furnaces with furnace temperature higher than 850℃. Third, epidemic-related medical waste can be transferred to the centralized disposal facilities for medical waste in neighboring areas for disposal according to the cross-regional coordination mechanism for emergency disposal. Fourth, infectious medical waste and other medical waste generated in the process of epidemic prevention and control shall be classified and managed separately. Fifth, facilitate medical institutions to adopt mobile medical waste disposal facilities for emergency disposal of epidemic-related medical waste and exempt them from environmental impact assessment and other procedures.

7.4 How to manage and dispose of medical waste in rural areas?

Rural areas that are not able to carry out centralized disposal of medical waste shall dispose of the medical waste generated by them on their own on site. Self-disposal of medical waste shall meet the following basic requirements: the used disposable medical instruments and medical waste easy to cause injury shall be disinfected and destroyed; medical waste that can be burned shall be burned in time; medical waste that cannot be burned shall be put together for landfill after disinfection.

7.5 What are the special management requirements for the qualification of medical waste disposal facilities during an epidemic?

The *Technical Guidelines on the Emergency Disposal of Medical*

Waste from the Treatment of COVID-19 (Trial) states that medical institutions that use movable medical waste disposal facilities to dispose of medical waste on their own or in neighboring medical institutions can be exempted from environmental impact assessment, medical waste business license and other procedures, but shall reasonably set up disposal sites, avoid environmentally sensitive areas such as protection areas for drinking water source and concentrated residential areas, and they shall report to the municipal authorities of health and ecological environment.

7.6 How shall areas with insufficient medical waste disposal capacity manage and dispose of medical waste during an epidemic?

The *Management Plan for Medical Waste in Response to Influenza A (H1N1) Epidemic* issued by the former Ministry of Environmental Protection of China on May 18, 2008 states that areas with insufficient disposal capacity for medical waste shall dispose of medical waste according to the emergency plan; for areas that indeed have no capability for centralized disposal of medical waste due to special reasons, especially remote rural areas, medical and health institutions can carry out on-site incineration of medical waste.

The *Technical Guidelines on the Emergency Disposal of Medical Waste from the Treatment of COVID-19 (Trial)* issued by the Ministry of Ecology and Environment of the people's Republic of China on

January 28, 2020 states that cities with gaps in medical waste disposal capacity shall coordinate other cities to establish a cross-regional coordination mechanism for emergency disposal. If centralized disposal indeed cannot be carried out due to special reasons, medical waste can be incinerated on site according to the plan determined by the local people's government.

7.7 How shall medical waste disposal be reported during the coronavirus (COVID-19) epidemic?

Local competent departments of ecology and environment administration at all levels shall report according to requirements of the Ministry of Ecology and Environment and local people's government so as to report medical waste disposal information in a timely manner.

Centralized medical waste disposal units shall regularly report the collection and centralized disposal of medical waste to local ecology and environment authorities above the county level, daily or weekly according to specific situation and needs.

8 Management Requirements for Medical Institutions during an Epidemic

8.1 How shall the used masks be treated during an epidemic?

The *Guidelines for the Protection of People at Different Risks of*

Infection from COVID-19 issued by the National Health Committee states that masks used by healthy people shall be treated according to the classification requirements for household garbage. Masks worn by suspected patients or confirmed patients shall not be discarded at random and shall be regarded as medical waste and treated in strict accordance with relevant procedures of medical waste.

8.2 During an epidemic, is the household garbage at the initial diagnosis isolation point a kind of medical waste?

Notification on Strengthening the Medical Management of the Initial Diagnosis Isolation Point of Novel Coronavirus Pneumonia issued by the National Health Committee: the initial diagnosis isolation point is designated by the local government, outside medical institutions, and is used to treat suspected COVID-19 patients with mild symptoms. It can be hotels, guesthouses and inns with certain conditions. Waste generated from initial diagnosis isolation points during the diagnosis and treatment, including medical waste and household garbage, shall be managed as medical waste in accordance with the *Notification on the Management of Medical Waste in Medical Institutions during the Novel Coronavirus Pneumonia Epidemic*.

8.3 What are the special requirements for classified collection of medical waste during an epidemic?

Regarding the collection of COVID-19-related medical waste,

the *Notification on the Management of Medical Waste in Medical Institutions during the Novel Coronavirus Pneumonia Epidemic* further requires that: the collection bin for medical waste shall be equipped with a foot pedal and a cover. When classifying and collecting used disposable isolation gowns and protective suits, squeezing is strictly prohibited.

The *Technical Guidelines on the Emergency Disposal of Medical Waste from the Treatment of COVID-19 (Trial)* clearly states that: infectious medical waste generated during the prevention and treatment of pneumonia epidemic shall be disinfected, packaged in strict accordance with the *Standard of Packaging Bags, Containers and Warning Marks Specific to Medical Waste*, and then placed in designated turnover bins (boxes) or disposable special packaging containers.

8.4 What are the special requirements for medical waste packaging during an epidemic?

Regarding the packaging of COVID-19-related medical waste, the *Notification on the Management of Medical Waste in Medical Institutions during the Novel Coronavirus Pneumonia Epidemic* further requires that: double-layer packaging bags shall be used to contain medical waste. Gooseneck-type sealing shall be adopted and each layer shall be sealed separately. When the outer surface of the packaging bag and sharps box containing medical waste is contaminated by infectious waste, an extra layer of packaging bag shall be added. The special note on the label of each packaging bag and sharp box shall be marked with "新型冠状病毒感染的肺炎" ("Novel Coronavirus Pneumonia" in English) or

abbreviated as “新冠” (“Coronavirus” in English). Medical waste generated in potentially contaminated areas and contaminated areas of fever clinics and impatient areas (wards) for COVID-19 patients and suspected COVID-19 patients shall be disinfected by spraying (uniformly spraying needed) 1,000 mg/L chlorine-containing disinfectant or be covered with a layer of medical waste packaging bag before leaving the contaminated areas.

8.5 What are the special requirements for medical waste to be transported in medical institutions during an epidemic?

The *Notification on the Management of Medical Waste in Medical Institutions during the Novel Coronavirus Pneumonia Epidemic* requires that: when transporting medical waste, transport staff shall prevent the special packaging bags and sharps boxes containing medical waste from damage, keep off the medical waste from directly contacting their bodies, and avoid the leakage and spread of medical waste. At the end of each day's transport, transport tools shall be cleaned and disinfected with 1,000 mg/L chlorine-containing disinfectant. If the transport tools are contaminated by infectious medical waste, they shall be disinfected in time.

8.6 What are the special requirements for the temporary storage of medical waste during an epidemic?

The *Notification on the Management of Medical Waste in Medical*

Institutions during the Novel Coronavirus Pneumonia Epidemic requires that: medical and health institutions with capabilities can store infectious medical waste generated during the prevention and treatment of pneumonia epidemic in a temporary storage place at a special site and under the management of a special person, and shall not mixed it with other medical wastes and household garbage.

During the COVID-19 epidemic, the temporary storage time of medical waste in medical and health institutions shall not exceed 24 hours.

The temporary storage places of medical waste in medical and health institutions shall be disinfected by specifically-designated person by spraying the wall or mopping the floor with 0.2% ～ 0.5% peracetic acid or 1,000 mg/L to 2,000 mg/L chlorine-containing disinfectant, once every morning and once every afternoon.

8.7 What are the special requirements for medical waste registration during an epidemic?

The *Notification on the Management of Medical Waste in Medical Institutions during the Novel Coronavirus Pneumonia Epidemic* requires that: layer-by-layer registration and transfer shall be carried out between medical waste generation department, transport personnel, staff at the temporary storage place and transfer personnel of the medical waste disposal unit and it shall be clarified that the medical waste is from COVID-19 patients or suspected COVID-19 patients.

The *Notification on the Management of Medical Waste in Medical Institutions during the Novel Coronavirus Pneumonia Epidemic* requires that: in areas where conditions permit, fixed special vehicles can be arranged to transport infectious medical waste generated during the prevention and treatment of pneumonia epidemic separately; duplicate tables for the transfer shall be filled in separately from other medical wastes; and a standing book shall be established.

8.8 What are the requirements for temporary hospitals to build temporary storage warehouses during an epidemic?

Construction shall be carried out in accordance with the requirements for the temporary storage warehouse for medical waste in infectious disease hospitals, and the construction standards shall not be lowered due to the temporality. Main requirements include: it shall be away from medical areas, food processing areas, personnel activity areas and household garbage storage places, facilitate the entry and exit of medical waste transporter, transport tools and vehicles, be equipped with strict sealing measures, be managed by special (part-time) staff to prevent non-staff from contacting medical waste, be equipped with measures against rats, mosquitoes, flies and cockroaches, be equipped with measures against leakage and rain erosion, be convenient to clean and disinfect, avoid direct sunlight, and have obvious warning symbols for hazardous waste, warning symbols for medical waste and warning

symbols writing "禁止吸烟饮食" ("No Smoking or Eating" in English). For temporary storage of pathological waste, low-temperature storage or anti-corrosion conditions shall be provided.

9 Management Requirements for Medical Waste Disposal Units during an Epidemic

9.1 What are the special requirements for the collection and transport of medical waste during an epidemic?

Infectious medical waste generated in the prevention and treatment process of the COVID-19 epidemic shall be transported by special medical waste transport vehicles or by vehicles temporarily re-purposed according to the requirements for medical waste transport vehicles; plan special driving routes to avoid densely populated areas and transport time shall avoid rush hours. "新型冠状病毒感染的肺炎" ("Novel Coronavirus Pneumonia" in English) or "新冠" ("COVID-19" in English) shall be specially noted.

9.2 What exemptions can be provided to facilitate emergency disposal of COVID-19-related medical waste?

The *Technical Guidelines on the Emergency Disposal of Medical Waste from the Treatment of COVID-19 (Trial)* states that: medical

institutions that use mobile medical waste disposal facilities to dispose of medical waste on their own or in neighboring medical institutions can be exempted from environmental impact assessment, medical waste business license and other procedures, but shall reasonably set up disposal sites, avoid environmentally sensitive areas such as protection areas for drinking water source and concentrated residential areas, and they shall report to the municipal health and ecological environment authorities. The supplier of mobile medical waste disposal facilities shall ensure that the disposal effect of medical waste meets the requirements of relevant standards and technical specifications.

9.3 What are the special requirements for the storage time limit of medical waste in centralized disposal units during an epidemic?

During the COVID-19 epidemic, medical waste delivered to the disposal site shall be disposed of as soon as possible and the temporary storage time in the disposal unit shall not exceed 12 hours.

9.4 Is it necessary to set up an isolation area for medical waste disposal during an epidemic?

The disposal unit must set up an isolation area for medical waste disposal. The isolation zone shall be clearly marked and irrelevant personnel are not allowed to enter; the isolation area of the disposal unit must be under the responsibility of special personnel, and shall be disinfected by spraying the wall, floor, and surface of articles or by

mopping the floor according to the methods and frequencies required by competent health department.

9.5 What are the disinfection requirements for medical waste disposal sites during an epidemic?

The isolation area for medical waste disposal must be managed by special personnel responsible for disinfection by spraying the wall, floor, and surface of articles or by mopping the floor with 0.2% ~ 0.5% peracetic acid or 1,000 ~ 2,000 mg/L chlorine-containing disinfectant, once every morning and once every afternoon.

9.6 What are the special requirements for occupational health protection during an epidemic?

During an epidemic, daily temperature monitoring of operators shall be strengthened. Where conditions permit, front-line operators for medical waste collection, storage, transport and disposal can be arranged to live together. Medical and health waste treatment/disposal personnel shall wash their hands and disinfect themselves immediately after the transport or disposal operation is completed, and personnel returning from special areas shall take a bath; before leaving the work area, the staff must disinfect the used tools with chlorine-containing disinfectant.

9.7 How to choose the disposal method for medical waste generated by COVID-19 patients?

Medical waste generated by COVID-19 patients can be disposed

of by high-temperature incineration, or non-incineration methods such as high-temperature steam disinfection, microwave disinfection and chemical disinfection, while ensuring the disposal effect.

9.8 What are the technical requirements and matters needing attention for emergency disposal of medical waste by non-professional disposal facilities?

In order to carry out emergency disposal of the COVID-19-related medical waste by non-professional disposal facilities for medical waste such as hazardous waste incineration facilities, household garbage incineration facilities and industrial furnaces, first, it is necessary to carry out health management and epidemic prevention in accordance with the requirements of the competent health department; second, special unloading and receiving areas as well as cleaning and disinfection areas shall be designated for medical waste, necessary rainproof and leak-proof measures shall be added, and special driving routes shall be planned for medical waste transport vehicles and managed by special personnel; third, warning symbols shall be set up at medical waste receiving site and restriction measures shall be carried out; fourth, special feeding equipment is recommended so as to prevent secondary cross-contamination caused by contact between medical waste and other incineration materials; fifth, attention shall be paid to the compatibility of medical waste and other incineration materials to ensure stable and controllable operation of the processing equipment; sixth, technical

operators shall receive necessary technical training.

10 Management and Disposal Cases during the COVID-19 Epidemic

10.1 A case of emergency disposal of medical waste using household garbage incineration facilities

In order to cope with emergency situations such as shutdown of medical waste incineration facilities for overhaul and public health emergencies, in early 2014, with the support of Shanghai Municipal Bureau of Ecology and Environment, Shanghai Municipal Health Commission, and Shanghai Landscaping & City Appearance Administrative Bureau, Shanghai Chengtou Group Corporation developed the *Shanghai Medical Waste Emergency Disposal Guarantee Plan*, completed the construction of supporting facilities such as loading and unloading area for medical waste turnover box, cleaning and disinfection area, stacking area and sewage collection in the pretreatment workshop of Laogang General Industrial Solid Waste Landfill Site, and finished the construction of supporting facilities such as medical waste unloading area, loading elevator, medical waste feeding device and sanitation and epidemic prevention control in Laogang Renewable Energy Utilization Center. After the emergency disposal facilities were checked and accepted by experts, the first emergency disposal of medical waste using household garbage incineration facilities in Shanghai was started in February 2014. From 2014 to 2019, the emergency disposal

volume of medical waste in Shanghai municipal household garbage incineration facilities was 850 tons, 1,930 tons, 4,630 tons, 7,560 tons, 9,310 tons and 14,490 tons respectively. In the emergency disposal process, health and epidemic prevention control have been carried out in accordance with the *Shanghai Municipal Health Management Code for Medical Waste and the Disinfection Methods for Supplies Related to Medical Waste Management*. The hourly proportion of mixed incineration of medical waste and household garbage has been controlled within 5%. Pollutant emission index is subject to the *Standard for Pollution Control on the Municipal Solid Waste Incineration* (DB 31/768—2013).

10.2 A case of disposal of medical waste using cement kiln co-processing technology

After the outbreak of the COVID-19 epidemic, several factories under a cement group entered an emergency disposal state, earnestly testing related equipment and conducting personnel operation drills, ready to receive and dispose of medical waste at any time. At the beginning of February, the enterprise received the task of emergency disposal of medical waste. The Group and its disposal factories immediately recalled relevant technicians and operators, set up isolation areas, and orderly carried out preparations such as deployment of protective materials and personnel drills. Every link including disinfection, collection and transport, delivery and disposal in the

treatment process has been carefully arranged. During medical waste disposal, its subordinate environmental companies arranged several fully sealed, leak-proof container horizontal type dumpers equipped with GPS to undertake medical waste transport tasks. Each dumper is equipped with a pace car to ensure safe transport. At the same time, there is no transfer point or hold-up in the disposal process in order to dispose of the waste generated within the same day, which not only ensures the safety of employees and the safety and efficiency of the disposal process, but also completely eliminates the risk of secondary pollution of medical waste. From February 4 to March 6, totally 55 tons of medical waste was disposed of.

10.3 Using mobile emergency disposal system for medical waste (Case 1)

An enterprise with a mobile disposal system for medical waste participated in the emergency treatment of epidemic-related medical waste in Wuhan City, Hubei Province, responsible for the emergency disposal of medical waste in the First people's Hospital of Jiangxia District and taking into account Leishenshan Hospital at the same time. The system adopts skid-mounted high-temperature steam technology for medical waste disposal which can be fixed or vehicle-mounted, and it mainly consists of medical waste sterilizer, sterilization trolley, cooling circulating water auxiliary system, crusher, unloader, automatic vehicle access mechanism (or cylinder propulsion device), and elevator.

The equipment itself is equipped with a steam generator and an air compressor, meaning it needs not to be connected with an external steam source but only needs connection with power electricity and water sources, thus facilitating access to medical institutions in epidemic areas. Daily processing capacity of the equipment is 1.5 tons.

10.4 Using mobile emergency disposal system for medical waste (Case 2)

In order to fully support China's fight against the COVID-19 epidemic, the United Nations Industrial Development Organization and the Foreign Environmental Cooperation Center of the Ministry of Ecology and Environment of China have jointly developed and implemented an emergency response project for the COVID-19 epidemic in China. A set of mobile high-temperature steam equipment for medical waste emergency treatment with a disposal capacity of 1.8 tons/day supported by the project has been put into use in Hongshan District of Wuhan City. The service scope of the equipment covers all designated hospitals for COVID-19 in Hongshan District and is responsible for the treatment of high-risk and highly infectious medical waste, with daily treatment capacity of medical waste up to 1.8 tons. Considering the actual needs of the epidemic area and in combination with the technical experience of the previously completed Global Environment Facility (GEF) -*China Medical Waste Environmental Sustainable Management Project*, vehicle-mounted mobile high-temperature steam medical waste

treatment technology is adopted. This technology not only reduces the risk of infection during the long-distance transport of medical waste, but also effectively avoids dioxins generation and emissions into the air. It can not only provide mobile and on-site treatment services during an epidemic, but also supplement the standby treatment capacity during the overhaul of centralized treatment equipment, thus achieving the long-term goal of improving the emergency treatment capacity of medical waste in the epidemic area.

10.5 A case of medical waste transport under emergency situation

Wuhan Fire Department has participated in transporting medical waste. In order to ensure the safety during medical waste collection, transport and disposal in the designated treatment hospitals for COVID-19, the Epidemic Prevention Headquarters arranged fire protection department, environmental protection department and hospitals to jointly carry out the collection, transport and disposal of medical waste in accordance with relevant specifications. Wuhan Fire Prevention and Rescue Detachment supported 45 medical units in transporting waste.

10.6 How to make good use of mainstream media for publicity during an epidemic?

First, focus on publicity and guidance, and dynamically grasp the direction of public opinion on the management and disposal of

medical waste during the epidemic. According to requirements of the *Technical Guidelines on the Emergency Disposal of Medical Waste from the Treatment of COVID-19 (Trial)*, local competent departments of ecology and environment at all levels shall release relevant information according to requirements of the people's government at the corresponding level. The departments of ecology and environment within the jurisdiction shall actively communicate with mainstream media in a timely manner, strengthen publicity, release information without delay, regularly report progress and results of environmental protection work related to epidemic prevention and control, and vigorously publicize advanced typical examples on epidemic prevention and control in various aspects to promote positive energy. Second, make full use of the official website, Weibo, WeChat Official Account and other media platforms of the government and the department of ecology and environment to release regulatory newsletter of medical waste within the jurisdiction, so that the public can learn about the information in a timely manner; and actively contact those influential news media to timely publicize relevant knowledge of medical waste management to the public, introduce a series of epidemic prevention and control work carried out by the government, and popularize knowledge of epidemic prevention and control and environmental protection. Third, use big data platform to collect and master news and public opinions on the management, collection and disposal of medical waste, discover

negative news in time, and carry out quick response measures to solve outstanding problems reflected by the public and the media.